W9-ATJ-122

OVERCOMING MATH ANXIETY

OVERCOMING MATH ANXIETY

Sheila Tobias

W · W · NORTON & COMPANY · INC ·
NEW YORK

CREDITS: Pinhead Studios for the charts and illustrations.
Phoenix Mutual Life Insurance Co. for permission to use the photograph of the Phoenix Building in Hartford Connecticut.
Roland B Guay, for permission to use the three-dimensional test items (FIGS. IV-3 and IV-4) in Chapter Four from the *Purdue Spatial Visualization Test*, © 1976 by the Purdue Research Foundation. All rights reserved.
Educational Testing Service, Princeton New Jersey, for permission to use an item from the *Hidden Figures Test*, © 1962. All rights reserved.
Doubleday and Co. for permission to use material for the design of FIG. VI-10 in Chapter Six. taken from Lancelot Hogben, *Mathematics in the Making*, © 1960 Rathbone Books. United Feature Syndicate, Inc. and Charles Schulz for permission to use the Peanuts cartoons.

Copyright © 1978 by Sheila Tobias
Published simultaniously in Canada by George J. McLeod Limited, Toronto. Printed in the United States of America.
ALL RIGHTS RESERVED

Library of Congress Cataloging in Publication Data
Tobias, Sheila.
Overcoming math anxiety.
Includes bibliographical references.
1. Mathematics—Study and teaching—Psychological aspects. I title.
QA11.T67 510'.7 78–17583
ISBN 0–393–06439–5

2 3 4 5 6 7 8 9 0

To Paul and Rose Tobias
parents, mentors, and friends

Contents

Preface

If anyone had told me four years ago that I would be involved in any way with mathematics, I would have thought the idea absurd. Like many women of my generation, I managed to avoid math and science and even to avoid thinking about why I had driven these important and interesting fields out of my life. Until I was thirty years old, I never even dated a scientist, an engineer, or a math major. My math avoidance extended even to my social life.

Not that I was incompetent. I did well enough in sixth grade when word problems began to stump my friends. I mastered elementary and intermediate algebra and liked geometry. Math courses were especially attractive to an uncertain teenager because the subject produced right answers that no unfriendly teacher could take away from you. There was a kind of certainty in mathematics that I needed then, and I felt gratified when things came out right.

During high school, things changed. A powerful and subtle pressure went to work on me. For reasons I did not then understand, I began to feel more comfortable not doing math at all. In college and for years at work I persuaded myself that I was interested in ideas and in people, and I believed for a long time that measurements only interfered. Being verbal and quick to pick up concepts, I managed to get away with what I finally had to concede was a yawning gap in my understanding of the world.

In the mid-1960s I became a feminist and subscribed heartily to the notion that preference and better pay for technical work was just one more way to put women down. By the mid-1970s, however, I was troubled by the continuing occupational segregation of women. I then believed that women were being tracked into people-oriented and helping professions because such occupations seemed to them and their colleagues to be appropriate to their female roles. My analysis went something like this:

Traditionally only three roles are seen as appropriate for women: the mother, the wife, and the decorative or pleasing object. The elementary school teacher, nurse, and social worker are extensions of the mother role. The assistant-to, secretary, and lab technician are extensions of the wife role. The receptionist, stewardess, and public relations specialist are the decorative roles. My view seemed to be confirmed when, in the 1960s, stewardesses won the right to keep on working after the age of 35, defeating the airlines' argument that flight attendants must be young and attractive to do their jobs. And when a group of receptionists in a New York corporation threatened a "smile boycott" to demonstrate how important it was for them to be decorative, I felt my analysis was right on target.[1]

Although I realized by then that many of the choices women make are really made for them, I did not see for a very long time that women are predestined to study certain subjects and pursue certain occupations not only because these areas are "feminine" but because girls are socialized not to study math.

Then, in the midst of a busy professional life as a feminist and educator, I was caught up short one day in 1974 when an unheralded and as yet unpublished piece of information crossed my desk. Lucy Sells, a California sociologist, had done the simplest of surveys. She counted the years of high school math taken by freshmen males and females entering the University of California at Berkeley in 1972. Of the entering males, 57 percent had taken four years of high school math, but only 8 percent of the females had had the same amount of math preparation. Without four years of high school math, students at Berkeley were ineligible for the calculus sequence, unlikely to attempt chemistry or physics, and inadequately prepared for intermediate statistics and economics. Since they could not take the entry-level courses in these fields, 92 percent of the females would be excluded from ten out of twelve colleges at Berkeley and twenty-two out of forty-four majors. Instead, they would be restricted to the "feminine" fields: humanities, guidance and counselling, elementary education, foreign languages, and the fine arts. All other options were foreclosed even before these women arrived at Berkeley.[2]

Something clicked. My previous analysis that the attraction of the mother-like, wifelike, and decorative occupations entirely accounted for job segregation no longer seemed adequate. I began to wonder, instead, about the choices women make early on and about the significance of avoiding math.

Not many moments in a person's life can be captured in memory. We rarely stop in our tracks and consciously change direction. But I am certain now that when I looked up from Lucy Sells' report I was already contemplating math therapy as a way to get women over math avoidance. The possibilities excited me even more than the problem, for if we could get to the bottom of math avoidance and find a way to demystify mathematics, we might not only change the pattern of occupational segregation but challenge some longstanding notions about sex differences in overall ability. Even the most recent and comprehensive review of sex differences, published in 1974, still presented mathematics ability as sex-linked.[3] If we could demonstrate, instead, that incompetence in mathematics is learned in conformity to sex-role expectations, then we might be able to design effective compensatory programs for women and girls. And this might affect not only female choice and female performance but also female mental health.

And so in 1974 I began a campaign against math avoidance. The effort has taken me to other countries to observe the effects of differing expectations on young people and deeply into mathematics and mathematics teaching. I have examined the myths surrounding mathematics, tried out some intervention techniques in an experimental clinic at my university, conferred with others more familiar than I am with how and why people learn, and looked at the implications of math avoidance on American education and values.

This book contains the results of my inquiry. It is not principally a how-to book, except perhaps for Chapters Six and Seven, which demonstrate how mathematical ideas become more comprehensible when presented playfully and in words, not symbols. The book is mainly a discussion of how intimidation, myth, misunderstanding, and missed opportunities have affected a large proportion of the population. My principal purpose in writing this book is to convince women and men that their fear of mathematics is the result and not the cause of their negative experiences with mathematics, and to encourage them to give themselves one more chance.

Much of what I have to say stems from introspection for I am a victim as well as a reporter of math anxiety. Much is drawn from many hours spent with learners and practitioners of mathematics, trying to find out how they feel when they do mathematics. Feelings are, I believe, at the heart of the problem, although we are supposed to leave feelings outside the math classroom. I have also considered the debates surrounding mathematics learning and dealt with the issues of spatial visualization and cognitive styles. Many of my own insights and my overall conviction that math anxiety is curable come from three years observing a math clinic at work. I have seen undergraduates and adults, at first too fearful of math to elect a single quantitative course, to change jobs, or to ask for a promotion, alter their goals and, more important, their self-image as a result of some intervention. I have watched people learn to trust their own intuition again and to recapture the unfettered curiosity of their youth.

Four years ago, when I began, I hypothesized that mathematics anxiety and mathematics avoidance were feminist issues. Now I am not so sure. Observing men has shown me that some men as well as the majority of women have been denied the pleasures and the power that competence in math and science can provide.* The feminists sounded the alarm. But, as a result, people of both sexes are beginning to reassess their mathe-

*In this book, meant for people of both sexes, the pronouns "he," "she," "him," and "her" will be used interchangeably unless a particular point about males or females is being made.

matical potential. It is to give these people courage and direction that I have written this book.

References

[1]Sheila Tobias, "Occupational Segregation: The Next Great Hurdle," unpublished, 1973; and Louise Kapp Howe, *Pink Collar Workers,* New York, G. P. Putnam's Sons, 1977, *passim.*

[2]Lucy Sells' article was finally published in 1978 under the title, "Mathematics—a critical filter," *The Science Teacher*, Vol. 45 No. 2, Feb. 1978.

[3]E.E. Maccoby and C.N. Jacklin, *The Psychology of Sex Differences,* Stanford: Stanford University Press, 1974, *passim.*

Acknowledgments

Some will say that it is too early to write about math anxiety since the experts have only just begun to examine the nature and causes of math avoidance. I recognize that I will not have the last word on the subject. Nor can I take credit for having the first word. Long before I began to think about mathematics and mathematics avoidance, a number of researchers in education and mathematics had been studying underachievement among women and men in math. I could not have written this book at all without the previous work of Elizabeth Fennema, Julia Sherman, Lucy Sells, Lynn Fox, John Ernest, Mitchell Lazarus, Robert Davis, Peter Hilton, Lenore Blum, Alice Schafer and her colleagues at Wellesley College, Robert Rosenbaum and Lorelei Brush from Wesleyan University, and Stanley Kogelman, Michael Nelson, Dora Helen Skypek, Judith Jacobs and others. Their published work is cited in the reference sections following each chapter.

This listing of my predecessors does not imply that they agree with what I say. Some may find my evidence scanty and my speculations uncalled for. As my methods are humanistic, I am willing to speculate on the basis of what social scientists would consider inadequate evidence. That is because I am as interested in what is possibly true as in what is necessarily true and as concerned about the symptoms of math anxiety as about causes and correlations. I believe that only with careful scrutiny of the symptoms will we finally understand what it *feels* like to be poor at math.

I expect that my analysis of feelings may contain insights worth investigating with more care and I hope that the experts will acknowledge the value of my speculation just as I appreciate their greater caution.

Without support from the Fund for the Improvement of Post-Secondary Education (FIPSE) for the Math Clinic at Wesleyan and from the Sloan Foundation, which helped support the clinic in 1977–78, there would have been no project from which I could learn. A Ford Foundation grant gave me time off to work on problems of women in higher education in 1976–77 and the German Marshall Fund of the United States paid my expenses while I was observing math anxiety and avoidance in women in four European countries in the summer of 1976. Program officers are not sufficiently appreciated. People like Alison Bernstein of FIPSE, Mariam Chamberlain of the Ford Foundation, Arthur Singer of the Sloan Foundation, and Gerald Livingston of the German Marshall Fund take risks when they support interesting but unproven new ideas, which was all we had to offer them when we began to tackle math anxiety.

In the course of writing this book family and friends have become colleagues, and colleagues friends. I am most deeply indebted to Carlos Stern, who taught me to appreciate the power of mathematics without making me feel dumb. Writers do not usually cite conversations as sources of their material, but it was in countless conversations with him that I began to confront my own anxiety and avoidance of mathematics. Lucy

Knight, an education journalist, shares my deep commitment to demystifying all disciplines; from this perspective she commented upon the book throughout its writing. Persis Joan Herold gave me insight into how children learn and as a bonus sketched the drawings and charts. Mathematicians and mathematics educators Joel Schneider, Knowles Dougherty, Fred Fischer, and Susan Auslander patiently tried to straighten out my mathematics, and though they are not responsible for my errors, they get credit for what I finally do understand.

Bonnie Donady and Jean Smith of Wesleyan University, Jane S. Stein of Duke University, Joseph Warren and Stanley Kogelman of Mind over Math in New York, Deborah Hughes-Hallett of Harvard, and Charles Stuth and Beverly Prosser Gelwick of Stephens College in Missouri generously shared their experiences in running math courses and clinics. My father, Paul J. Tobias, taught me to think creatively about the word problems in Chapter Five, invented the letter division puzzles in Chapter Eight, and in his evaluation of my work continued the shaping of my mind that he began so many years ago.

There is no way to thank the hundreds of people who wrote to me after my article on math anxiety was published in *Ms* magazine in September 1976, but I drew heavily upon their math autobiographies; nor can I acknowledge all the people who regaled me with quarter-hour-long recollections about learning math when they met me somewhere and found out what I was doing. Some people can be thanked, however. Lorraine Karatkewicz, my administrative assistant and dear friend, made it possible for me to think while managing a clinic and fulfilling my obligations as Associate Provost. Colin

Campbell, president of Wesleyan, and an extraordinary Board of Trustees supported the project unstintingly from the beginning.

Last, I want to thank Robert Rosenbaum, professor of mathematics at Wesleyan and co-director of the Math Clinic, who has taught me more about learning than I can possible acknowledge here and who, at a critical moment in the planning of this project, said "Yes."

Middletown, Conn.
January 1978

OVERCOMING MATH ANXIETY

1

The Primacy of Mathematics, or If I Could Do Math I Would . . .

One day soon after we began working with people who avoid mathematics, a visiting Soviet mathematician stopped in at the Math Clinic. It took a while to explain to him what the clinic had been designed to do, how people who have long ago given up on their ability to learn math need to be recaptured and motivated to try again. Finally he seemed to understand. Then he began to laugh and with a big, generous smile he commented: "You Americans are all the same. You think everybody has to know everything."

In a way this is true, although he needn't have been quite so smug.* The United States graduates more young people from high school than most other nations,

*Soviet children seem to have just as much trouble learning elementary arithmetic as American children. A collection of essays on teaching arithmetic in the Soviet Union reports one child's strategy for getting the right answer as follows: "I add, subtract, multiply, and then divide until I get the answer that is in the back of the book."

and we have a larger proportion of people enrolled in some sort of post-secondary schooling than any other nation in the world. On paper at least we are committed to the belief that everyone should receive a general education and that specialized training should be postponed until later. Still, not everyone learns everything. Many of us elect to stay within a fairly narrow comfort zone of knowledge. If we do well at reading long, nonfiction books and writing essays about them, we take more courses in history. If we excel at learning one foreign language, we take another one or two. If we learn best by handling things, we opt for lab courses. And so on.

By the time we are adults we are pretty smart in our special fields but feel rather dumb when we move outside our comfort zone. What began as a set of preferences becomes in time a mental prison that makes us feel conflicted and even anxious about stepping outside. It is difficult enough for a young person to be a beginner, dependent on other people to show the way. To be a beginner as an adult takes a special kind of courage and enthusiasm for the subject. That is why I believe that a slight discomfort with mathematics acquired in elementary or secondary school can develop into a full-fledged syndrome of anxiety and avoidance by the time one has graduated from school and gone to work.

How do these preferences and antagonisms develop? Are they inevitable? Do we hate math, as so many say they do, because one elementary school teacher made us afraid of it? Did we simply get off on the wrong foot and get stuck there? Or have we accepted someone's assumption about our kind of mind and what we thought was the nature of mathematics? Were we too young, from a neurological point of view, when we were first exposed to numbers? Or was too much going on during our adolescent years when we should have been concentrating on algebra? Does the fault lie in the curriculum, in the subject, in ourselves, or in the society at large?

There are no easy answers to such questions. But some interesting insights have come from people who have avoided mathematics all their lives. In dozens of "math autobiographies," the nature of math anxiety, its causes, and its consequences begin to emerge.

"If I could do math I would . . ."*

Ask anyone who "hates math" to complete this sentence and the first response will be incredulity. "What, me do math? Impossible." If prodded, however, people reveal all kinds of unexplored needs. "If I could do math I would . . . fix my own car." "If I could do math I would . . . fly an airplane." "If I could do math I would work the F stops on my camera." "If I could do math . . ."

People who don't know what math is don't know what math isn't. Therefore, fear of math may lead them to avoid all manner of data and to feel uncomfortable working with things. Any mathematician will tell you that you don't need mathematics to work the F stops on the camera, or to fix the car, or even to start your own business.** But fearing math makes us wary of activities that *may* involve ratios (F stops), or neat rows of numbers (bookkeeping), or mechanics. What could we do if we could do math? It depends, of course, on how much

*"If I could do math I would" is a phrase used by Bonnie Donady of the Wesleyan Math Clinic in her counselling.

**F stops are a measure of the lens opening: The higher the number, the smaller the opening.

math we know and how confident we are of our knowledge. But it is safe to say that if we were comfortable with numbers we could do something we are not doing now.

Most people leave school as failures at math, or at least feeling like failures. Some students are not even given a chance to fail. Identified early as noncollege material, they are steered away from pre-college mathematics and tracked into business or "general math." But high school students who are not going on to college need algebra and geometry just as much as the college-bound. Algebra and geometry are not just pre-college courses. There is evidence that knowing these two subjects alone will make the difference between a low score and a high score on most standardized entry-level tests for civil service, federal service, and many industrial occupations, not to mention the armed services. Lucy Sells, who has been studying the link between mathematics and vocational opportunities, believes that knowledge of algebra and geometry divides the unskilled and clerical jobs from the better-paying, upwardly mobile positions available to high school graduates. She estimates that mastery of high school algebra alone will enable a high school graduate to do so much better on a civil service or industrial entrance exam that he or she could immediately start working at a higher level and be earmarked for on-the-job training as well. Just one more year of high school math could make the difference between a starting salary of $8,000 and one of $11,000.[1]

The influence of high school students' math programs on their future careers has also been recognized by Joseph Watson of the University of California at San Diego, who has written a brochure called "Parents,

Protect Your Children's Future: Get Them to Take Algebra and Geometry."[2] In this little document he spells out the importance of high school math in the job market. Beginning in 1978 the United States Civil Service Commission will collect detailed information about the relevance of high school math to test scores. It will ask everyone taking its standardized examinations to list how many math courses he has had. When this information is collected over the next two years we shall know what difference high school algebra and geometry make.

The terminal high school graduate is not the only one to suffer vocationally for lack of sufficient math. Just as poignant is the plight of the middle-aged food chain executive who complains that she cannot get any facts from the figures she encounters. Indeed, if she cannot interpret quantitative data in the form of balance statements, budgets, and records of sales, she is in trouble. As mathematician Lynn Osen puts it, business today needs people "who can understand a simple formula, read a graph and interpret a statement about probability."[3]

It comes as no surprise that fields like engineering, computer programming, accounting, geology, and other technical work all require some mathematics. It is not widely known, however, that mathematics, disguised as "quantitative analysis" or "handling data" or "planning," will be used even in nontechnical work areas. It can be very painful to discover in mid-career, like the food chain executive, that we cannot get facts from our figures or that an occupation that looked comfortably free of mathematics at the lower levels will require familiarity with quantitative methods for advancement. Yet those who supervise, plan, and manage in fields such as social work, librarianship, retail sales, school administration, and even publishing now need to be familiar with or at least willing to learn more math.

Thus, even the most thoughtfully prepared high school guidance material may be misleading. The information being made available to students even today tends to list only careers that explicitly require mathematics competency *at the very outset.* Often unmentioned is the new truth, as expressed in a recent speech by W. W. Sawyer: "The ability to think mathematically will have to become something taken for granted as much as the ability to read a newspaper is at present. Such a change will seem fantastic to some people. But so would universal literacy have seemed absurd a few centuries ago."

Since fear of math spills over into fear of "data" and "things," one way to measure the effect of math avoidance on our lives is to ask ourselves how we feel about "data" and "things." The Dictionary of Occupational Titles, published regularly by the U.S. Department of Labor, uses "data," "people," and "things" as one way to categorize occupations. A quick glance at any page in that volume indicates that if we cross "data" and "things" off our lists of possibilities, our occupational choices will be very limited indeed. Another measure of math avoidance can be inferred from vocational interest tests. Typically these group an individual's personal traits into categories which in turn can be translated into jobs. One of the tests, the Self-Directed Search Inventory, lists as "interest-types" *realistic, investigative, artistic, enterprising, social,* and *conventional.*[4] John Holland, designer of this inventory, also publishes an *Oc-*

cupation Finder in which occupations are coded according to the same categories. Engineering, for example, is defined as "realistic-investigative," management as "realistic-enterprising-social," and so on. In fact, far more occupations require realistic, enterprising, and investigative skills than require primarily "social" attributes.[5]

Of course, the characteristics of each occupation are derived from people currently at work. Thus, there is a tautology: if engineering is dominated by "realistic-investigative" types, people in the field will of course report that this configuration of skills and temperament is highly valued in engineering and essential to it. Also, these opinions may contain hidden biases against women and "feminine" traits. John Holland is himself very mindful of this and tries to factor out such bias. But when an entire society associates math, science, things, and data with male pursuits, it is difficult to eliminate this bias altogether. And nothing changes the fact that math avoidance is extremely limiting for people at all levels of work. Competence in math is, as Sells put it, truly a vocational filter.

Not everyone who avoided mathematics in school remains incompetent, however. In two years of interviewing adults who are math anxious and those who are not, I have found that some people learn to cope with numbers in interesting and sometimes extraordinary ways. They know they have gaps in their knowledge, but they are sufficiently confident and familiar with the way their own minds work to reconstruct problems and solve them.

In a session for a mixed group of adults, Knowles Dougherty, a specialist in teaching the math disabled, demonstrated this phenomenon. He began by asking the group to do a simple subtraction problem in their heads:

> A woman is 38 years old. It is now 1976.
> In what year was she born?

"Don't tell me the answer," he said. "But as I go around the room, do tell me how you worked out the solution." The methods were astonishingly varied. One adult male said, "I had that certain feeling I had to get to the nearest ten. So I added two to 38 to get 40 and then subtracted 40 from 76 and then added two to the answer." Another person reported that she had adjusted the problem to her own age. Starting with her own year of birth, she added and subtracted until she got 1938 (the correct answer).

Each adult was a little ashamed of his system. Everyone assumed that just as there was only one right answer to math problems, there was probably only one right way to subtract. But Dougherty reassured them that the systems they were using were legitimate algorithms (an algorithm being just a system for getting an answer). If you have to get to the nearest ten, then get there. If you have to use some personal reference point, use it. The fact that most people had not used the method they had been taught in school indicated that they had probably learned a fair amount of mathematics on their own since second grade.

Many adults know all this intuitively and develop ways to work out number problems that make sense to them. In Washington, D. C. a professional woman receives bimonthly account records from her bookkeeper. She can make no sense at all of what they show until she goes off by herself and reorganizes the data. She turns columns into rows and rows into boxes, adjusting the numerical information to fit her thinking style. Then she returns to the discussion with her bookkeeper, facts in hand. Some people need to draw pictures. Some have to speak the numbers or the problem out loud.

These people share a willingness to restructure a problem so that it makes sense to them. The executive who could not get the facts from her figures had not yet learned how to do that.

Why can't everyone make such a healthy accommodation? Why can't we just go out and learn the math we need when we need it? The rest of this book will deal in depth with what people remember about studying math and why they believe they cannot learn more. Apart from the psychological blocks that may grow up over time, some of the issues are these: Math is difficult because it is rigorous and complex. As we advance in math, the notation becomes abstract and general. This adds to its mysteriousness. Besides, there are conflicts between the common everyday use of words and the use of these words in math. Moreover, unless we are concurrently studying science or engineering, math is not as integrated into the rest of the curriculum as are reading, writing, and spelling. Hence we do not immediately apply the math we learn.

Math typically is not learned as part of a group effort. Since we work by ourselves, the process by which we either get or do not get an insight is very obscure. And since we do not learn how our own minds work, we never understand why one idea is difficult for us and the next one not.

Finally, people remember math as being taught in an atmosphere of tension created by the emphasis on right answers and especially by the demands of timed tests.

Their math teachers, who may well have been patient and sincere to begin with, became frustrated by the challenge of getting students to understand "simple" ideas; even if they did not start out this way, many of them eventually became cantankerous, short of patience, and contemptuous of error. Math anxious adults can recall with appalling accuracy the exact wording of a trick question or the day they had to stand at the blackboard alone, even if these events took place thirty years before.

It will surprise no one that millions do not learn mathematics successfully, however well or poorly it is taught. The most recent attempt to improve the teaching of arithmetic took place fifteen years ago. The New Math curriculum was developed with the best intentions by some of the finest mathematicians and math educators in the country. It was intended to correct the previous emphasis on memory and drill and to provide, instead, the "why" of operations.

Perhaps the fairest and most optimistic expectation of the New Math was stated by Francis Keppel, then Commissioner of Education, in a foreword written in 1963 to one visionary new look at the mathematics curriculum. Keppel said that the effort was intended to represent mathematics as the scholar himself regarded it, "complete with its sense of adventure, its unsolved questions and its groping toward the future."[6]

Indeed the goals were exciting: the curricular material developed was meant to bring a student over a period of thirteen years to an understanding of the structure of numbers, a readiness for calculus, familiarity with probability theory, and mastery of analytic geometry. The way this was to be accomplished was to eliminate both unnecessary drill and "real-life" problems, and in their place to teach more sophisticated math. The designers of the revised curricula were aware that with greater options for electives in junior and senior high school, fewer and fewer students would take advanced mathematics. Because of this and because they thought it was possible to teach more advanced math to younger people, they decided to concentrate on the lower grades.

True, the vocabulary of arithmetic had not been revised for a long time. A 1909 text I own, *The Complete Arithmetic,* is very much like the book I used in the 1940s.[7] Except perhaps for the chapter on "Problems of the Farm," which the teacher in my urban school may have skipped, the chapter headings, the sequence, the language, and even the problems seem very familiar. The design of the New Math included a new vocabulary. Although today experts in the field explain that pupils were not meant to learn by heart words like "commutative," "associative," and "distributive," these terms were taught in the lower grades along with "sets," "transformations," and "functions." To familiarize elementary school teachers with the new pedagogy, numerous clinics and retraining programs were established in the 1960s. Texts were rewritten, parents were urged to take a crash course in the New Math to learn to help their children with their homework, and, eventually, the reform spread to Europe as well.

Anyone who brought up a child in this period knows one thing for sure: the New Math helped aggravate the generation gap. If the parent had not taken advanced mathematics in college and was too anxious or indifferent to study a parents' guide to the New Math, the child would get no help at home. Mysterious as math had always been, it became even more so. How can you help a child who talks about "sets" when you have never heard of them and the child says *you* don't know what the teacher is doing? Students' unsatisfactory math performance today may be the result of this lack of help at home.

It is rumored that the New Math is on its way out. Whether or not this is true, many elementary school teachers no longer teach it, though their texts and lesson plans still pay it lip service. Even some college mathematics professors are changing their minds about its benefits. They are discouraged by the lack of manipulative skill and mechanical accuracy in today's college students. "Too much theory and not enough practice," is one criticism. "Too little 'guesstimating' " is another.

Yet, looking back on the New Math, I share the excitement and the fervor of its progenitors. The idea of a "spiral curriculum," alighting on the same ideas several times at higher and higher levels of complexity; the elimination of drill; the "discovery" approach; the development of intuition are experiences I would have savored, then and even now. Perhaps this is the issue: most pupils may simply have been too young and inexperienced to handle arithmetic presented in this fashion. On the other hand, adults who want to relearn arithmetic at a higher level might respond very well to it.

Why should adults want to relearn arithmetic? Why should those of us who have successfully avoided mathematics all these years bother about it at all? Mathematicians have usually been admired for their intelligence but also somewhat scorned for their otherworldliness. Why is their subject suddenly so central to modern life?

To some extent mathematics has always played a role, particularly at the higher occupational levels. The news is that math is increasingly pervasive and also that these occupational levels are beginning to be reached by different people. No longer are academically bright males from advantaged homes the only ones who are expected to go as far in math as they can. In the last twenty years, females and males from less advantaged backgrounds have been encouraged to take school seriously and to expect advancement on the job. These groups have traditionally had little success with higher math and have "mercifully" been allowed to drop it. Much to their surprise and dismay, as their careers advance, math returns to haunt them.[8]

The primacy of mathematics in today's world, then, exhibits the traditional themes of American history: ever-increasing dependence on technology and ever-increasing democratization of jobs.

As in every other aspect of American life, however, World War II accelerated changes that were already underway in the running of business and government. A typical story, probably neither entirely true nor entirely false, recounts that when Robert McNamara returned from his wartime assignment in logistics and supplies, he sold himself and his entire team to the Ford Motor Company with a promise to bring "systems analysis" to management. He and his colleagues had developed computer systems to keep track of the whereabouts of war material. McNamara persuaded Ford that systems analysis, then a very new idea, could give management the same kind of control over production, marketing, and distribution. Previously, except for scientific and engineering industries, the only people who generally used mathematics at work were those in

FIG. I-1

investment and finance or accounting and budgeting. Decisions elsewhere were made either by extrapolating from laboriously hand-processed data or by what is now recalled as "flying by the seat of one's pants". Although some amount of intuition always comes into decision making, the mix of data and intuition was significantly altered by systems analysis.

The benefit of systems analysis, which made McNamara so attractive to the Ford Motor Company, was a precise picture of what was actually going on in the business. Thus the best, or what was soon to be called the "optimal" business decision could be made. Before, it had been simply too difficult and too time-consuming to find out, for example, whether a particular department was paying its way by computing its gross sales against its share of operating costs. Operating costs could not be precisely assigned to each product made. Management might keep a division going just because it *looked* profitable.

Systems analysis was useful not only in evaluation and assessment but also in making decisions. With so many square yards of tin and so many orders for big cans, medium-sized cans, and small cans, each at a certain price and each providing a certain percentage of profit, what combination of large, medium-sized, and small cans should a firm manufacture in a month, a week, or a day in order to maximize profit? Assisted by the computer, which could handle the calculations quickly and without error, business entered a new era of decision making. Decisions about when to buy and how much, when to sell and at what price, when to unload because keeping something might get too expensive, when to reorder, when to expand, when to contract, and even when to go out of business, were all easier to make with mathematics than without.

While systems analysis was revolutionizing business practices, the methods of assessing nonprofit, socially useful programs were becoming quantitative, too. Before the war, a technique called benefit-cost analysis was hardly known outside the water resources area of the federal government. To get some sense of the usefulness of water projects for flood control, irrigation, recreation, and production of electric power, the Congress required that large projects be evaluated in terms of their benefits in relation to their costs. Ideally, the benefits of a dam, for example, should always exceed its costs by some amount. (Thus the phrase "benefit-cost ratio" came into being.)

There are, of course, difficulties in comparing such benefits as irrigation, wilderness recreation, and hydroelectric power and these conflicting values gave rise to serious debates for years. But benefit-cost analysis remained attractive because it provided one way to evaluate high-cost projects in terms of one another. Since the Congress usually voted these projects in an omnibus bill, including many other projects that could not be debated individually, the benefit-cost ratio might be the only feature of a project that would be noted when the time came to vote.

Just as benefit-cost analysis was developed to meet a problem in the water resources area, so was it borrowed to meet similar problems in other areas as the federal government began spending large sums for other kinds of domestic projects.

Today benefit-cost and its first cousin "cost-effectiveness" are basic tools of program evaluation. They are used in assessing the war on poverty, welfare reform, health measures, and even educational programs. And this means that the math avoider cannot escape these techniques even if she flees from the business world to the nonprofit sector. However difficult it is to quantify the value of a bird or the good reputation of an institution, administrators today want to know at least whether a proposed project will cost more than it benefits, even if they decide to ignore these figures for the sake of some larger (or smaller) goal.

Since even in the nonprofit sector projects are always

competing for funds, it is important to know their cost-effectiveness: which project will get more impact, dollar for dollar? Will a dollar spent on infant nutrition have a greater payoff in terms of poor children's intellectual performance than a dollar spent on a Head Start program in an urban school? Will a dollar spent in oil exploration produce more benefits than a dollar spent in home insulation? Measuring impact and comparing the cost-effectiveness of programs has become as necessary a part of social work and prison reform as assessing the costs and benefits of automation has become part of running a large library.

But even more has changed than the worlds of business and nonprofit enterprises. At the University of California at Berkeley, twenty-two out of forty-four majors require the calculus sequence or a course in intermediate statistics.[9] This reflects a change in the nature of research as many of the social sciences embrace techniques for taking large surveys, for handling the masses of data that result from these surveys, and for relating the phenomena these data reveal. Regression analysis, perhaps the most useful of all the statistical techniques being applied to social issues today, allows us to compare answers to several questions in a large, many-faceted study, to find out whether the answers are related in any way; whether age, for example, correlates with a negative opinion on abortion rights; or whether religion, sex, and geographical location are more powerful correlates than age; whether smoking correlates with the incidence of throat cancer; whether College Board scores predict college performance.

In a study of energy conservation, for example, two factors are obviously related to the amount of fuel a family consumes: the weather and the size of the apartment or house. But other factors, such as the price of fuel, whether the renter or the landlord pays for the fuel, whether there is thermostatic control in every room, are also significant. How is one to figure out which of these is most important and which are interdependent?

Since many problems, like this one, involve more than two factors, and since "multiple" regression analysis beyond three dimensions is virtually impossible without a computer, the common use of these statistical techniques has been growing with the availability of computers and the refinement of computer programming. The computer can do what no human brain could do in a reasonable amount of time: sort and resort large amounts of data seeking patterns and connections among the elements.

This emphasis on statistical techniques should not of course diminish the importance of human judgment. On the contrary, such techniques make human judgment all the more critical, since wrong questions will produce useless if not dangerously misleading answers. As the computer specialists put it, "Garbage in, garbage out." Still, frequent use of all these techniques in modern research means that the math avoider will be excluded from these career areas. One graduate school admissions dean put it very bluntly when she said, "I'd rather have a math major enter psychology at the graduate level than take in a psychology major who had studied no math."

Most sophisticated and powerful of all the new techniques are the large mathematical models that reproduce mathematically a system, a process, or an entire institution, incorporating information so complex that nobody could possibly retain it all in any number of files. These models are designed to simulate the real world so that the researcher or manager can try out some idea, such as a shift in price per item or in temperature per beaker, and "see" what effect the change will have on the entire process. Such a model, for example, is being developed in Cambridge, Massachusetts for the U.S. Department of Transportation. It will evaluate AMTRAK's punctuality and suggest ways to minimize lateness. The mathematician in charge of the model has all the arrival and departure information for all stations in a given period. He also has the number of passengers who get on and off in each place. His job is to find a way to ascertain not simply how many AMTRAK trains were how late (this would be simple) but how many people were inconvenienced by how much by AMTRAK lateness.

The mathematician's first job is to create a basic unit for measuring passenger inconvenience, such as a unit of passenger-late-minutes-per-mile. This whole process may sound very exotic, but if AMTRAK has only enough surplus money one year to fix so many miles of track or to add only so many extra ticket agents or to reschedule only so many stops, the question for management will be, "What combination of these efforts will minimize total passenger inconvenience and maximize passenger satisfaction?" Information on the passenger-late-minutes-per-mile for any one segment of track might help.

What is the net gain to a college if it raises tuition by a certain amount? On the one hand, a number of parents will decide not to send their children to that college. Also, the amount of financial aid given to students will have to be increased to meet the higher tuition rate (which will reduce the net gain to the college). In addition, the cost of recruiting students will increase. To predict the effect of this decision, it would be useful to be able to simulate the situation with a model that reflects the revenues, the expenses, the public relations impact, the loss of some proportion of applicants, and all the other consequences of the action. Of course, some decision will be made without such a model. The college may simply call up a neighboring comparable institution and ask if *it* is planning to raise tuition this year, then let this information be its guide.

Is it better to manufacture many of one item in one plant distant from some of the markets or to distribute the manufacturing to various plants? Cost of transportation in this case has to be weighed against the savings made when large amounts are manufactured at one time in one place. A mathematical model that includes all the details of the cost of manufacturing and transportation, including variations by season and by size of order, can answer this question in minutes.

Apart from illustrating the power of mathematics in solving practical problems, these examples also make it clear that, contrary to our grade-school notion that all of mathematics produces exact right answers, mathematics seems to be a process of organizing information into categories. Thus math is quite capable of dealing both with qualitative data and with uncertainty. No wonder familiarity with some of these techniques is useful at the higher reaches of almost every occupation.

Long ago in Europe, anyone who wanted to participate in government, in centers of learning, and even in certain trades, had to learn Latin and sometimes even Greek. Quantitative methods, if not mathematics itself, have become the Latin of the modern era. The vocabulary that derives from systems analysis and the kind of dynamic interactions described in these examples can best be understood by people who have studied math at the level of calculus and beyond. Whether one approves or disapproves of these developments, one has to concede that literacy has a new dimension: mathematical competence.

References

[1]Lucy Sells, "High School Math as a Vocational Filter for Women and Minorities," unpublished article, *passim.,* 1974; and further conversations with the author.
[2]Joseph Watson, "Parents, Protect Your Children's Future: Get Them to Take Algebra and Geometry," San Diego, the Partnership Program in conjunction with The Third College, University of California at San Diego, n.d.
[3]Lynn Osen, "The Feminine Math-tique," Pittsburgh, K.N.O.W., 1971, *passim.*
[4]John Holland, "Self-Directed Search Inventory," Palo Alto, Consulting Psychologist Press, 1970.
[5]John Holland, *Occupation Finder,* Palo Alto, Consulting Psychologist Press, 1970.
[6]Francis Keppel, Foreword, *Goals for School Mathematics,* The Report of the Cambridge Conference on School Mathematics, Educational Services, Inc., Boston, Houghton Mifflin, 1963, p. vii.
[7]Wentworth and Smith, *The Complete Arithmetic,* Boston, Athenaeum Press, 1909.
[8]Knowles Dougherty, adviser and commentator on this section, suggested the duality of the postwar developments: not just the mathematicization through systems analysis which I had identified but also the higher-level jobs to which less advantaged groups were aspiring.
[9]Sells, *op. cit., passim.*

2

The Nature of Math Anxiety: Mapping the Terrain

A warm man never knows how a cold man feels.
—Alexander Solzhenitsyn

Symptoms of Math Anxiety

The first thing people remember about failing at math is that it felt like sudden death. Whether it happened while learning word problems in sixth grade, coping with equations in high school, or first confronting calculus and statistics in college, failure was sudden and very frightening. An idea or a new operation was not just difficult, it was impossible! And instead of asking questions or taking the lesson slowly, assuming that in a month or so they would be able to digest it, people remember the feeling, as certain as it was sudden, that they would *never* go any further in mathematics. If we assume, as we must, that the curriculum was reasonable and that the new idea was merely the next in a series of learnable concepts, that feeling of utter defeat was simply not rational; and in fact, the autobiographies of math anxious college students and adults reveal that no matter how much the teacher reassured them, they sensed that from that moment on, as far as math was concerned, they were through.

The sameness of that sudden death experience is evident in the very metaphors people use to describe it. Whether it occurred in elementary school, high school, or college, victims felt that a curtain had been drawn, one they would never see behind; or that there was an impenetrable wall ahead; or that they were at the edge of a cliff, ready to fall off. The most extreme reaction came from a math graduate student. Beginning her dissertation research, she suddenly felt that not only could she never solve her research problem (not unusual in higher mathematics), but that she had never understood advanced math at all. She, too, felt her failure as sudden death.

Paranoia comes quickly on the heels of the anxiety attack. "Everyone knows," the victim believes, "that I don't understand this. The teacher knows. Friends know. I'd better not make it worse by asking questions. Then everyone will find out how dumb I really am." This paranoid reaction is particularly disabling because fear of exposure keeps us from constructive action. We feel guilty and ashamed, not only because our minds seem to have deserted us but because we believe that our failure to comprehend this one new idea is proof that we have been "faking math" for years.

In a fine analysis of mathophobia, Mitchell Lazarus explains why we feel like frauds. Math failure, he says, passes through a "latency stage" before becoming obvious either to our teachers or to us. It may in fact take some time for us to realize that we have been left behind. Lazarus outlines the plight of the high school student who has always relied on the memorize-what-to-do approach. "Because his grades have been satisfactory, his problem may not be apparent to anyone, including himself. But when his grades finally drop, as they must, even his teachers are unlikely to realize that his problem is not something new, but has been in the making for years."[1]

It is not hard to figure out why failure to understand mathematics can be hidden for so long. Math is usually taught in discrete bits by teachers who were themselves taught this way; students are tested, bit by bit, as they go along. Some of us never get a chance to integrate all these pieces of information, or even to realize what we are not able to do. We are aware of a lack, but though the problem has been building up for years, the first time we are asked to use our knowledge in a new way, it feels like sudden death. It is not so easy to explain, however, why we take such personal responsibility for having "cheated" our teachers and why so many of us believe that we are frauds. Would we feel the same way if we were floored by irregular verbs in French?

One thing that may contribute to a student's passivity is a common myth about mathematical ability. Most of us believe that people either have or do not have a mathematical mind. It may well be that mathematical imagination and some kind of special intuitive grasp of mathematical principles are needed for advanced research, but surely people who can do college-level work in other subjects should be able to do college-level math as well. Rates of learning may vary. Competence under time pressure may differ. Certainly low self-esteem will interfere. But is there any evidence that a student needs to have a mathematical mind in order to succeed at *learning math?*

Leaving aside for the moment the sources of this myth, consider its effects. Since only a few people are supposed to have this mathematical mind, part of our passive reaction to difficulties in learning mathematics is that we suspect we may not be one of "them" and are waiting for our nonmathematical mind to be exposed. It is only a matter of time before our limit will be reached, so there is not much point in our being methodical or in attending to detail. We are grateful when we survive fractions, word problems, or geometry. If that certain moment of failure hasn't struck yet, then it is only temporarily postponed.

Sometimes the math teacher contributes to this myth. If the teacher claims an entirely happy history of learning mathematics, she may contribute to the idea that some people—specifically her—are gifted in mathematics and others—the students—are not. A good teacher, to allay this myth, brings in the scratch paper he used in working out the problem to share with the class the many false starts he had to make before solving it.

Parents, especially parents of girls, often expect their children to be nonmathematical. If the parents are poor at math, they had their own sudden death experience; if math was easy for them, they do not know how it feels to be slow. In either case, they will unwittingly foster the idea that a mathematical mind is something one either has or does not have.

Interestingly, the myth is peculiar to math. A teacher of history, for example, is not very likely to tell students that they write poor exams or do badly on papers because they do not have a historical mind. Although we might say that some people have a "feel" for history, the notion that one is *either* historical or nonhistorical is patently absurd. Yet, because even the experts still do not know how mathematics is learned, we tend to think of math ability as mystical and to attribute the talent for it to genetic factors. This belief, though undemonstrable, is very clearly communicated to us all.

These considerations help explain why failure to comprehend a difficult concept may seem like sudden death. We were kept alive so long only by good fortune. Since we were never truly mathematical, we had to memorize things we could not understand, and by memorizing we got through. Since we obviously do not have a mathematical mind, we will make no progress, ever. Our act is over. The curtain down.

Ambiguity, Real and Imagined

> What is a satisfactory definition? For the philosopher or the scholar, a definition is satisfactory if it applies to those things and only those things that are being defined; this is what logic demands. But in teaching, this will not do: a definition is satisfactory only if the students understand it.
>
> —H. Poincaré

Mathematics autobiographies show that for the beginning student the language of mathematics is full of ambiguity. Though mathematics is supposed to have a very precise language, more precise than our everyday use (this is why math uses symbols), it is true that mathematical terms are never wholly free of the connotations we bring to words, and these layers of meaning may get in the way. The problem is not that there is anything wrong with math; it is that we are not properly initiated into its vocabulary and rules of grammar.

Some math disabled adults will remember, after fifteen to thirty years, that the word "multiply" as used for fractions never made sense to them. "Multiply," they remember wistfully, always meant "to increase." That is the way the word was used in the Bible, in other contexts, and surely the way it worked with whole numbers. (Three times six always produced something larger than either three or six.) But with fractions (except the improper fractions), multiplication always results in something of smaller value. One-third times one-fourth equals one-twelfth, and one-twelfth is considerably smaller than either one-third or one-fourth.

Many words like "multiply" mean one thing (like "increase rapidly") when first introduced. But in the larger context (in this case all rational numbers), the apparently simple meaning becomes confusing. Since students are not warned that "multiplying" has very different effects on fractions, they find themselves searching among the meanings of the word to find out what to do. Simple logic, corresponding to the words they know and trust, seems not to apply.

A related difficulty for many math anxious people is the word "of" as applied to fractions. In general usage, "of" can imply division, as in "a portion of." Yet, with fractions, one-third *of* one-fourth requires multiplication. We can only remember this by suspending our prior associations with the word "of," or by memorizing the rule. Or, take the word "cancel" as used carelessly with fractions. We are told to "cancel" numerators and denominators of fractions. Yet nothing is being "cancelled" in the sense of being removed for all time. The same holds true for negative numbers. Once we have learned to associate the minus sign with subtraction, it takes an explicit lesson to unlearn the old meaning of minus; or, as a mathematician would put it, to learn its meaning as applied to a new kind of number.

Knowles Dougherty, a skilled teacher of mathematics, notes:

> It is no wonder that children have trouble learning arithmetic. If you ask an obedient child in first grade, "What is Zero," the child will call out loudly and with certainty, "Zero is nothing." By third grade, he had better have memorized that "Zero is a place-holder." And by fifth grade, if he believes that zero is a number that can be added, subtracted, multiplied by and divided by, he is in for trouble.[2]

People also recall having problems with shapes, never being sure for example whether the word "circle" meant the line around the circle or the space within. Students who had such difficulties felt they were just dumber than everyone else, but in fact the word "circle" needs a far more precise definition. It is in fact neither the circumference nor the area but rather "the locus of points in the plane equidistant from a center" (Fig. II-1).*

A mind that is bothered by ambiguity—actual or perceived—is not usually a weak mind, but a strong one. This point is important because mathematicians argue that it is not the subject that is fuzzy but the learner who is imprecise. This may be, but as mathematics is often taught to amateurs differences in meaning be-

*One student, learning to find the "least common denominator," took the phrase "least common" to mean "most unusual" and hunted around for the "most unusual denominator" she could find. Instead of finding the smallest common denominator, then, she found a very large one and was appropriately chastised by her teacher for misunderstanding the question.

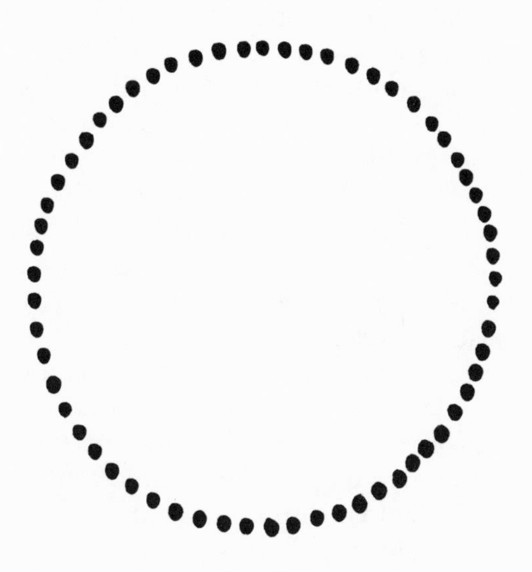

FIG. II-1

tween common language and mathematical language need to be discussed. Besides, even if mathematical language is unambiguous, there is no way into it except through our spoken language, in which words are loaded with content and associations. We cannot help but think "increase" when we hear the word "multiply" because of all the other times we have used that word. We have been coloring circles for years before we get to one we have to measure. No wonder we are unsure of what "circle" means. People who do a little better in mathematics than the rest of us are not as bothered by all this. We shall consider the possible reasons for this later on.

Meanwhile, the mathematicians withhold information. Mathematicians depend heavily upon customary notation. They have a prior association with almost every letter in the Roman and Greek alphabets, which they don't always tell us about. We think that our teachers are choosing X or a or delta (Δ) arbitrarily. Not so. Ever since Descartes, the letters at the end of the alphabet have been used to designate unknowns, the letters at the beginning of the alphabet usually to signify constants, and in math, economics, and physics generally Δ means "change" or "difference." Though these symbols appear to us to be chosen randomly, the letters are loaded with meaning for "them."

In more advanced algebra, the student's search for meaning is made even more difficult because it is almost impossible to visualize complex mathematical relationships. For me, the fateful moment struck when I was confronted by an operation I could neither visualize nor translate into meaningful words. The expression $X^{-2}=\frac{1}{X^2}$ did me in. I had dutifully learned that exponents such as ^{2} and ^{3} were shorthand notations for multiplication: a number or a letter squared or cubed was simply multiplied by itself twice or three times. Trying to translate math into words, I considered the possibility that X^{-2} meant something like "X not multiplied by itself" or "multiplied by not-itself." What words or images could convey to me what X^{-2} really meant? To all these questions—and I have asked them many times since—the answer is that $X^{-2}=\frac{1}{X^2}$ is a definition consistent with what has gone before. I have been shown several demonstrations that this definition is indeed consistent with what has gone before.* But at the time

*While interviewing for this book, I have finally found out that negative two is a different kind of number from positive two and that it was naïve of me to think that it would have the same or similar effect on X. And it does work. If you divide X^3 by X^5 (remember you subtract exponents when you divide) you end up with $\frac{1}{X^2}$. See the following:

$$X^{-2} \equiv \frac{X^3}{X^5} = \frac{XXX}{XXXXX} = \frac{1}{XX} = \frac{1}{X^2}$$

I did not want a demonstration or a proof. I wanted an explanation!

I dwell on the X^{-2} example because I have often asked competent mathematicians to recall for me how they felt the first time they were told $X^{-2} = \frac{1}{x^2}$. Many remember merely believing what they were told in math class, or that they soon found the equivalency useful. Unlike me, they were satisfied with a proof and an illustration that the system works. Why some people should be more distrustful about such matters and less willing to play games of internal consistency than others is a question we shall return to later.

Willing suspension of disbelief is a phrase that comes not from mathematics or science but from literature. A reader must give the narrator an opportunity to create images and associations and to "enter" these into our mind (the way we "enter" information into a computer) in order to carry us along in the story or poem. The very student who can accept the symbolic use of language in poetry where "birds are hushed by the moon," or the disorienting treatment of time in books by Thomas Mann and James Joyce, may balk when mathematics employs familiar words in an unfamiliar way. If willingness to suspend disbelief is specific to some tasks and not to others, perhaps it is related to trust. One counsellor explains math phobia by saying, "If you don't feel safe, you won't take risks." People who don't trust math may be too wary of math to take risks.

A person's ability to accept the counter-intuitive use of time in Thomas Mann's work and not the new meaning of the negative exponent does not imply that there are two kinds of minds, the verbal and the mathematical. I do not subscribe to the simple-minded notion that we are one or the other and that ability in one area leads inevitably to disability in the other. Rather, I think that verbal people feel comfortable with language early in life, perhaps because they enjoyed success at talking and reading. When mathematics contradicts assumptions acquired in other subjects, such people need special reassurance before they will venture on.

Conflicts between mathematical language and common language may also account for students' distrust of their intuition. If several associated meanings are floating around in someone's head and the text considers only one, the learner will, at the very least, feel alone. Until someone tries to get inside the learner's head or the learner figures out a way to search among the various meanings of the word for the one that is called for, communication will break down, too. This problem is not unique to mathematics, but when people already feel insecure about math, linguistic confusion increases their sense of being out of control. And so long as teachers continue to argue, as they have to me, that words like "multiply" and "of," the negative exponents, and the "circles" or "disks" are not ambiguous at all but perfectly consistent with their definitions, then students will continue to feel that math is simply not for them.

Some mathematics texts solve the problem of ambiguity by virtually eliminating language. College-level math textbooks are even more laconic than elementary texts. One reason may be the difficulty of expressing mathematical ideas in language that is easily agreed upon. Another is the assumption that by the time students get to college they should be able to read symbols. But for some number of students (we cannot know how many since they do not take college-level math) proofs, symbolic formulations, and examples are not enough. After I had finally learned that X^{-2} must equal $\frac{1}{X^2}$ because it was consistent with the rule that when dividing numbers with exponents we subtract the exponents, I looked up "negative exponents" in a new high school algebra text. There I found the following paragraph.

Negative and Zero Exponents

The set of numbers used as exponents in our discussion so far has been the set of positive integers. This is the only set which can be used when exponents are defined as they were in Chapter One. In this section, however, we would like to expand this set to include all integers (positive, negative and zero) as exponents. This will, of course, require further definitions. These new definitions must be consistent with the system and we will expect all of the laws of exponents as well as all previously known facts to still be true.[3]

Although this paragraph is very clear in setting the stage to explain negative exponents through definitions which are presumably forthcoming, it does not provide a lot of explanation. No wonder people who need words to make sense of things give up.

The Dropped Stitch

"The day they introduced fractions, I had the measles." Or the teacher was out for a month, the family moved, there were more snow days that year than ever before (or since). People who use events like these to account for their failure at math did, nevertheless, learn how to spell. True, math is especially cumulative. A missing link can damage understanding much as a dropped stitch ruins a knitted sleeve. But being sick or in transit or just too far behind to learn the next new idea is not reason enough for doing poorly at math forever after. It is unlikely that one missing link can abort the whole process of learning elementary arithmetic.

In fact, mathematical ideas that are rather difficult to learn at age seven or eight are much easier to comprehend one, two, or five years later if we try again. As we grow older, our facility with language improves; we have many more mathematical concepts in our minds, developed from everyday living; we can ask more and better questions. Why, then, do we let ourselves remain permanently ignorant of fractions or decimals or graphs? Something more is at work than a missed class.

It is of course comforting to have an excuse for doing poorly at math, better than having to concede that one does not have a mathematical mind. Still, the dropped stitch concept is often used by math anxious people to excuse their failure. It does not explain, however, why in later years they did not take the trouble to unravel the sweater and pick up where they left off.

Say they did try a review book. Chances are it would not be helpful. Few texts on arithmetic are written for adults.* How insulting to go back to a "Run, Sport, run!" level of elementary arithmetic, when arithmetic can be infinitely clearer and more interesting if it is discussed at an adult level.

Moreover, when most of us learned math we learned

*Deborah Hughes-Hallett is writing a book (W.W. Norton, 1978) for adults and college students that starts with arithmetic and brings the reader up to calculus, in two volumes.

dependence as well. We needed the teacher to explain, the textbook to drill us, the back of the book to tell us the right answers. Many people say that they never mastered the multiplication table, but I have encountered only one person so far who carries a multiplication table in his wallet. He may have no more skills than the others, but at least he is trying to make himself autonomous. The greatest value of using simple calculators in elementary school may, in the end, be to free pupils from dependence on something or someone beyond their control.

Adults can easily pick up those dropped stitches once they decide to do something about them. In one math counselling session for educators and psychologists, the following arithmetic bugbears were exposed:

How do you get a percentage out of a fraction like $7/16$?

Where does "pi" come from?

How do you do a problem like: Two men are painters. Each paints a room in a different time. How long does it take them to paint the room together?*

The issues were taken care of within half an hour.

This leads me to believe that people are anxious not because they dropped a stitch long ago but rather because they accepted an ideology that we must reject: *that if we haven't learned something so far it is probably because we can't.*

*See Chapter Six for a discussion of fractions and percents; see Chapter Five for a discussion of the Painting-the-Room Problem. *Pi* can be derived by drawing many-sided polygons (like squares, pentagons, hexagons, etc.) and measuring the ratio between their diameters and their perimeters. Even if you do this roughly, the ratios will approach 3.14.

Fear of Being Too Dumb or Too Smart

© 1966 United Features Syndicate, Inc.

One of the reasons we did not ask enough questions when we were younger is that many of us were caught in a double bind between a fear of appearing too dumb in class and a fear of being too smart. Why anyone should be afraid of being too smart in math is hard to understand except for the prevailing notion that math whizzes are not normal. Boys who want to be popular can be hurt by this label. But it is even more difficult for girls to be smart in math. Matina Horner, in her survey of high-achieving college women's attitudes toward academic success, found that such women are especially nervous about competing with men on what they think of as men's turf.[4] Since many people perceive ability in mathematics as unfeminine, fear of success may well interfere with ability to learn math.

The young woman who is frightened of seeming too smart in math must be very careful about asking questions in class because she never knows when a question is a really good one. "My nightmare," one woman remembers, "was that one day in math class I would innocently ask a question and the teacher would say, 'Now that's a fascinating issue, one that mathematicians spent years trying to figure out.' And if that happened, I would surely have had to leave town, because my social life would have been ruined." This is an extreme case, probably exaggerated, but the feeling is typical. Mathematical precocity, asking interesting questions, meant risking exposure as someone unlike the rest of the gang.

It is not even so difficult to ask questions that gave the ancients trouble. When we remember that the Greeks had no notation for multi-digit numbers and that even Newton, the inventor of the calculus, would have been hard pressed to solve some of the equations given to beginning calculus students today, we can appreciate that young woman's trauma.

At the same time, a student who is too inhibited to ask questions may never get the clarification needed to go on. We will never know how many students developed fear of math and loss of self-confidence because they could not ask questions in class. But the math anxious often refer to this kind of inhibition. In one case, a counsellor in a math clinic spent almost a semester persuading a student to ask her math teacher a question *after* class. She was a middling math student, with a B in linear algebra. She asked questions in her other courses, but could not or would not ask them in math. She did not entirely understand her inhibition, but with the aid of the counsellor, she came to believe it had something to do with a fear of appearing too smart.

There is much more to be said about women and mathematics. The subject will be discussed in detail in Chapter Three. At this point it is enough to note that some teachers and most pupils of both sexes believe that boys naturally do better in math than girls. Even bright girls believe this. When boys fail a math quiz their excuse is that they did not work hard enough. Girls who fail are three times more likely to attribute their lack of success to the belief that they "simply cannot do math."[5] Ironically, fear of being too smart may lead to such passivity in math class that eventually these girls also develop a feeling that they are dumb. It may also be that these women are not as low in self-esteem as they seem, but by failing at mathematics they resolve a conflict between the need to be competent and the need to be liked. The important thing is that until young women are encouraged to believe that they have the right to be smart in mathematics, no amount of supportive, nurturant teaching is likely to make much difference.

Distrust of Intuition

> Mathematicians use intuition, conjecture and guesswork all the time except when they are in the classroom.
>
> —Joseph Warren, Mathematician

Thou shalt not guess.
—Sign in a high school math classroom

At the Math Clinic at Wesleyan University, there is always a word problem to be solved. As soon as one is solved, another is put in its place. Everyone who walks into the clinic, whether a teacher, a math anxious person, a staff member, or just a visitor, has to give the word problem a try. Thus, we have stimulated numerous experiences with a variety of word problems and by debriefing *both* people who have solved these problems and people who have given up on them, we gain another insight into the nature of math anxiety.

One of the arithmetic word problems that was on the board for a long time is the Tire Problem:

A car goes 20,000 miles on a long trip. To save wear, the five tires are rotated regularly. How many miles will each tire have gone by the end of the trip?

Most people readily acknowledge that a car has five tires and that four are in use at any one time. Poor math students who are not anxious or blocked will poke around at the problem for a while and then come up with the idea that four-fifths of 20,000, which is 16,000 miles, is the answer. They don't always know exactly why they decided to take four-fifths of 20,000. They sometimes say it "came" to them as they were thinking about the tires on the car and the tire in the trunk. The important thing is that they *tried* it and when it resulted in 16,000 miles, they gave 16,000 a "reasonableness test." Since 16,000 seemed reasonable (that is, less than 20,000 miles but not a whole lot less), they were pretty sure they were right.

The math anxious student responds very differently.

FIG. II-2

The problem is beyond her (or him). She cannot begin to fathom the information. She cannot even imagine how the five tires are used (See FIG. II-2.) She cannot come up with any strategy for solving it. She gives up. Later in the debriefing session, the counsellor may ask whether the fraction four-fifths occurred to her at all while she was thinking about the problem. Sometimes the answer will be yes. But if she is asked why she did not try out four-fifths of 20,000 (the only other number in the problem), the response will be—and we have heard this often enough to take it very seriously—"I figured that if it was in my head it had to be wrong."

The assumption that if it is in one's head it has to be wrong or, as others put it, "If it's easy for me, it can't be math," is a revealing statement about the self. Math anxious people seem to have little or no faith in their own intuition. If an idea comes into their heads or a strategy appears to them in a flash they will assume it is wrong. They do not trust their intuition. Either they remember the "right formula" immediately or they give up.

Mathematicians, on the other hand, trust their intuition in solving problems and readily admit that without it they would not be able to do much mathematics. The difference in attitude toward intuition, then, seems to be another tangible distinction between the math anxious and people who do well in math.

The distrust of intuition gives the math counsellor a place to begin to ask questions: Why does intuition appear to us to be untrustworthy? When has it failed us in the past? How might we improve our intuitive grasp of mathematical principles? Has anyone ever tried to "educate" our intuition, improve our repertoire of ideas by teaching us strategies for solving problems? Math anxious people usually reply that intuition was not allowed as a tool in problem solving. Only the rational, computational parts of their brain belonged in math class. If a teacher or parent used intuition at all in solving problems he rarely admitted it, and when the student on occasion did guess right in class he was punished for not being able to reconstruct his method. Yet people who trust their intuition do not see it as "irrational" or "emotional" at all. They perceive intuition as flashes of insight into the rational mind. Victims of math anxiety need to understand this, too.

The Confinement of Exact Answers

> "Computation involves going from a question to an answer. Mathematics involves going from an answer to a question."
>
> —Peter Hilton, Mathematician

Another source of self-distrust is that mathematics is taught as an exact science. There is pressure to get an exact right answer, and when things do not turn out right, we panic. Yet people who regularly use mathematics in their work say that it is far more useful to be able to answer the question, "What is a little more than five multiplied by a little less than three?" than to know *only* that five times three equals 15.* Many math anxious adults recall with horror the timed tests they were subjected to in elementary, junior and senior high school with the emphasis on getting a unique right answer. They liked social studies and English better because there were so many "right answers," not just one. Others were frustrated at not being able to have discussions in math class. Somewhere they or their teachers got the wrong notion that there is an inherent contradiction between rigor and debate.

This emphasis on right answers has many psychological benefits. It provides a way to do our own evaluation on the spot and to be judged fairly whether or not the teacher likes us. Emphasis on the right answer, however, may result in panic when that answer is not at

*A little more than five multiplied by a little less than three will produce a range between 12.5 and 16.5. Inequalities, of which this is one example, are common in more advanced math, as are equations that have more than one solution.

hand and, even worse, lead to "premature closure" when it is. Consider the student who does get the right answer quickly and directly. If she closes the book and does not continue to reflect on the problem, she will not find other ways of solving it, and she will miss an opportunity to add to her array of problem-solving methods. In any case, getting the right answer does not necessarily imply that one has grasped the full significance of the problem. Thus, the right-answer emphasis may inhibit the learning potential of good students and poor students alike.

In altering the learning atmosphere for the math anxious the the tutor or counsellor needs to talk frankly about the difficulties of doing math. The tutor's scratch paper might be more useful to the students than a perfectly conceived solution. Doing problems afresh in class at the risk of making errors publicly can also link the tutor with the student in the process of discovery. Inviting all students to put their answers, right or wrong, before the class will relieve some of the panic that comes when students fail to get the answer the teacher wants. And, as most teachers know, looking carefully at wrong answers can give them good clues to what is going on in students' heads.

Although an answer that checks can provide immediate positive feedback, which aids in learning, the right answer may come to signify authoritarianism (on the part of the teacher), competitiveness (with other students), and painful evaluation. None of these unpleasant experiences is usually intended, any more than the premature closure or panic, but for some students who are insecure about mathematics the right-answer emphasis breeds hostility as well as anxiety. Worst of all, the "right answer" isn't always the right one at all. It is only "right" in the context of the amount of mathematics one has learned so far. First graders, who are working only with whole numbers, are told they are "right" if they answer that five (apples) cannot be divided between two (friends). But later, when they work with fractions, they will find out that five *can* be equally divided by giving each friend two and one-half apples. In fact both answers are right. You cannot divide five one-dollar bills equally between two people without getting change.

The search for the right answer soon evolves into the search for the right formula. Some students cannot even put their minds to a complex problem or play with it for a while because they assume they are expected to know something they have forgotten.

Take this problem for, example,

Amy Lowell goes out to buy cigars. She has 25 coins in her pocket, $7.15 in all. She has seven more dimes than nickels and she has quarters, too. How many dimes, nickels, and quarters does she have?

Most people who have done well in high school algebra will begin to call the number of nickels X, the number of dimes $X + 7$, and the number of quarters $7.15 minus $(5X + 10X + 70)$ without realizing that Amy Lowell must have miscounted her change, because even if all 25 coins in her pocket were quarters (the largest coin she has), her change would total $6.25, not $7.15.*

This is a tricky problem, which is fair, as opposed to a trick problem which is not. But it also shows how searching for the right formula can cause us to miss an

*I am indebted to Jean Smith for this example.

obviously impossible situation. The right formula may become a substitute for thinking, just as the right answer may replace consideration of other possibilities. Somehow students of math should learn that the power of mathematics lies not only in exactness but in the processing of information.

Self-Defeating Self-Talk

One way to show people what is going on in their heads is to have them keep a "math diary," a running commentary of their thoughts, both mathematical and emotional, as they do their homework or go about their daily lives. Sometimes a tape recorder can be used to get at the same thing. The goal is twofold: to show the student and the instructor the recurring mathematical errors that are getting in the way and to make the student hear his own "self-talk." "Self-talk" is what we say to ourselves when we are in trouble. Do we egg ourselves on with encouragement and suggestions? Or do we engage in self-defeating behaviors that only make things worse?

Inability to handle frustration contributes to math anxiety. When a math anxious person sees that a problem is not going to be easy to solve, he tends to quit right away, believing that no amount of time or rereading or reformulation of the problem will make it any clearer. Freezing and quitting may be as much the result of destructive self-talk as of unfamiliarity with the problem. If we think we have no strategy with which to begin work, we may never find one. But if we can talk ourselves into feeling comfortable and secure, we may let in a good idea.

To find out how much we are talking ourselves into failure we have to begin to listen to ourselves doing math. The tape recorder, the math diary, the self-monitoring that some people can do silently are all techniques for tuning in to ourselves. Most of us who handle frustration very poorly in math handle it very well in other subjects. It is useful to watch ourselves doing other things. What do we do there to keep going? How can these strategies be applied to math?

At the very minimum this kind of tuning in may identify the particular issue giving trouble. It is not very helpful to know that "math makes me feel nervous and uncomfortable" or that "numbers make me feel uneasy and confused," as some people say. But it may be quite useful to realize that one kind of problem is more threatening than another. One excerpt from a math diary is a case in point:

> Here I go again. I am always ready to give up when the equation looks as though it's too complicated to come out right. But the other week, an equation that started out looking like this one did turn out to be right, so I shouldn't be so depressed about it.

This is constructive self-talk. By keeping a diary or talking into a tape recorder we can begin to recognize our own pattern of resistance and with luck we may soon learn to control it. This particular person is beginning to understand how and why she jumps to negative conclusions about her work. She is learning to sort out the factual mistakes she makes from the logical and even the psychological errors. Soon she will be able to recognize the mistakes she makes *only* because she is anxious. Note that she has been encouraged to think and to talk about her feelings while doing mathematics. She is not ashamed or guilty about the most irrational of thoughts, not frightened to observe even the onset of depression in herself; she seems confident that her mind will not desert her.

The diary or tape recorder technique has only been tried so far with college-age students and adults. So far as we can tell, it is effective only when used in combination with other nonthreatening teaching devices, such as acceptance of discussion of feelings in class, psychological support outside of class, and an instructor willing to demystify mathematics. The goal in such a situation is not to get the right answer. The goal is to achieve mastery and above all autonomy in doing math. In the end, we can only learn when we feel in control.

References

This chapter is based primarily on interviews with and observations of math anxious students and adults. These people are not typical of those who are math incompetent. Most of them are very bright and enjoy school success in other subjects, but they avoid or openly fear mathematics.

[1]Mitchell Lazarus, "Mathophobia: Some Personal Speculations," *The Principal,* January/February, 1974, p. 18.
[2]Knowles Dougherty. Personal communication to the author.
[3]J. Louis Nanney and John L. Cable, *Elementary Algebra: A Skills Approach,* Boston, Allyn and Bacon, Inc., 1974, p. 215.
[4]Matina Horner, "Fear of Success," *Psychology Today,* November, 1969, p. 38ff.
[5]Sanford Dornbusch, as quoted in John Ernest, "Mathematics and Sex," *American Mathematics Monthly,* Vol. 83, No. 8, October, 1976, p. 599.

3

*Mathematics and Sex**

Men are not free to avoid math; women are.

In a major address to the American Academy of Arts and Sciences in 1976, Gerard Piel, publisher of *Scientific American,* cited some of the indicators of mathematics avoidance among girls and young women. "The SAT record plainly suggests that men begin to be separated from women in high school," he noted. "At Andover [an elite private high school] 60 percent of the boys take extra courses in both mathematics and science, but only 25 percent of the girls. . . . By the time the presently graduating high school classes are applying to graduate school," he concluded, "only a tenth as many young women as men will have retained the confidence and capacity to apply to graduate study in the sciences."[1]

Some other measures of mathematics avoidance among females are these:

Girls accounts for 49 percent of the secondary school students in the United States but comprise only 20 percent of those taking math beyond geometry.

*"Mathematics and Sex" is the title chosen by John Ernest for his important essay on the problem. The essay was published by the *American Mathematical Monthly* and reprinted by the Ford Foundation in 1976.

The college and university population totals 45 percent women, yet only 15 percent of the majors in pure mathematics are women.

Women make up 47 percent of the labor force and 42 percent of those engaged in professional occupations. Yet they are only 12 percent of the scientific and technical personnel working in America today.

Are these data simply evidence of individual preference, or do they represent a pattern of math avoidance and even math anxiety among women? We know that there are differences in *interest* between the sexes. What we do not know is what causes such differences, that is whether these are differences in ability, differences in attitude, or both; and, even more important, whether such differences, if indeed they exist, are innate or learned.*

Most learning psychologists doing research today are environmentalists; that is, they tend to be on the "nurture" side of the nature-nurture controversy. Most of them would therefore not subscribe to the man on the street's belief that mathematics ability is just one of those innate differences between men and women that can neither be ignored nor explained away. Yet even the most recent research on sex differences in intelligence accepts the fact that performance in math varies by gender.[3] Because this is assumed to be natural and inevitable (if not genetic in origin) for a long time the causes of female underachievement in mathematics have not been considered a promising area for study

*This chapter draws particularly on the work of Elizabeth Fennema, Julia Sherman, and Lynn Fox, who were commissioned by the National Institute of Education in 1976 to review critically research on mathematics and women. Also important is the work of John Ernest, Lorelei Brush, and Lynn Osen. Michael Nelson has prepared an exhaustive annotated bibliography on all aspects of math ability and disability, including a chapter on mathematics and sex. See the reference section for details about these works.[2]

and certainly not an urgent one.

But recently, as women began to aspire to positions in fields previously dominated by men, this attitude began to change. The women's movement and the accompanying feminist critique of social psychology can be credited, I believe, with the rise in interest in mathematics and sex and with the formulation of some important new questions. Do girls do poorly in math because they are afraid that people (especially boys) will think them abnormal if they do well, or is it because girls are not taught to believe that they will ever need mathematics? Are there certain kinds of math that girls do better? Which kinds? At what ages? Are there different ways to explain key concepts of math that would help some girls understand them better?

One example of a new approach is the initially informal survey undertaken by John Ernest, a professor of mathematics at the University of California at Santa Barbara. In 1974, Ernest volunteered to teach a freshman seminar about Elementary Statistics. Believing that these notions are best learned in a concrete situation, he turned his seminar into an investigation of the relationships, real and imagined, between mathematics and sex. His students fanned out into neighboring junior and senior high schools to interview teachers and students about girls' and boys' performance in mathematics. The results of their inquiry were nearly always the same. Both boys and girls, they were told, have a fair amount of trouble doing math and most of them do not like the subject very much. The difference between them is this: boys stick with math because they feel their careers depend on it and because they have more confidence than girls in their ability to learn it.[4]

Ernest augmented the report with a survey of other

people's research on mathematics and women and sent it off in the fall of 1975 to be published by the *American Mathematics Monthly,* a journal read mostly by mathematicians. The article was, however, considered so important that the Ford Foundation pre-printed it as a brochure and mailed it to 44,000 educators around the country. Partly in response to the interest sparked by Ernest's article and partly to the spread of other new research on women and math, in summer 1976 the National Institute of Education commissioned three experts in the field to review critically all the research that had been done so far and to propose new research priorities for the agency.

In one of those reports, Elizabeth Fennema, a professor of curriculum and instruction at the University of Wisconsin-Madison, concluded, in agreement with Ernest's findings, that the problem may not be so much one of discrimination or of differences in ability, but rather one of *math avoidance* on the part of women and girls. Whatever the other reasons might be, not taking math in the eleventh and twelfth grades would surely affect females' math performance later on. Fennema writes: "The problem with girls is not the ability to learn math but the willingness to study math." She bases her conclusion on several recent studies of her own, including a close inspection of math enrollments in Wisconsin high schools. In one of these, at the twelfth grade level, 45 percent of the boys but only 29 percent of the girls were still taking math. Fennema suggests that if four years of high school math were required of all high school students, we would go far toward eliminating the problem of differences in math performance between the sexes.[5]

As Elizabeth Fennema well knows, this is not a realistic solution. Some number of girls (and boys) cannot be pressed or cajoled into taking more math. Still, her argument is worth bearing in mind as we proceed in this analysis of sex differences because girls might do as well as boys if they took the same amount of math. We do not know whether they would because several past national studies of math achievement showing lower scores for girls did not consider that girls take fewer courses in math than boys. A new study, the National Assessment of Performance and Participation of Women in Mathematics, begun in the fall of 1978, will attempt to correct this lack by noting carefully the amount and nature of math previously studied.[6]

In fact, math avoidance is not just a female phenomenon. Most people of both sexes stop taking math before their formal education is complete. Few people become mathematicians and many very smart people do not like math at all. Thus, "dropping out" of math is nearly universal, and is by no means restricted to girls and women. From this perspective, girls who avoid math and math-related subjects may simply be getting the message sooner than boys that math is unrewarding and irrelevant, but boys will also get that message in time.

A recent survey of attitudes toward math among ninth and twelfth graders demonstrated this point very well. Although ninth grade girls had a more negative attitude toward math than ninth grade boys, by the twelfth grade boys had caught up. The researcher concluded that by age 17 a majority of all students have developed an aversion to math, which is tragic but certainly not sex-related.[7]

What then is gender-related? What can we say with certainty about mathematics and sex?

Performance and "Ability"

Since innate "ability" can be measured only by performance on some test, those of us who are interested in sex differences in mathematics are forced to look at the results of tests given to boys and girls to measure their math achievement at different points in their lives. Not all of the tests have been of high quality, but some of the most widely quoted assessments have been carefully done and most have surveyed large, national populations. Yet, even when sex differences are found, they do not necessarily imply sex differences in "mathematical intelligence" or even in aptitude for math.

There are many reasons to be cautious. One is that tests of math performance, however well designed, are still tests. Therefore whatever people bring to a test—test anxiety, math anxiety, or hostility to math—will interfere with their performance. Thus even the best national assessments of girls' and boys' mathematical performance may not reveal as much as we think they do.

Take Project Talent, for example. Completed in 1960, it surveyed thousands of boys and girls. Yet, for all its care in sampling and testing, girls were probably compared to boys who had taken more years of math. It should come as no surprise that by grade twelve, males significantly outperformed females.

Even where greater care has been taken to compare boys and girls with similar mathematics backgrounds, the conclusions reached are not always qualified by other factors or warranted by the degree of difference found. The College Board and National Merit administrators report a consistent pattern of male superiority in math, but do not adjust for the fact that their populations may already be specializing in a field. In the National Assessment of Educational Progress, done in 1972–73, the researchers found some differences but far from enough to warrant their conclusion that the "advantage displayed by males, particularly at the older ages, is overwhelming."[8]

Besides, there is contrary evidence. When Elizabeth Fennema and Julia Sherman compared boys and girls in four Wisconsin high schools in 1974–75, they located disparities in math achievement in only two of the four schools and found only minimal differences at that. Taken alone, this finding is not significant. The important difference between their results and others' probably has to do with careful comparison of boys and girls having similar math backgrounds. It is also possible, as Fennema and Sherman point out, that sex differences in math performance are diminishing—hence, the more recent the research the fewer the differences. Nevertheless, their results must be dealt with.

Although it is tempting to draw conclusions from global comparisons of boys' and girls' performance on math achievement tests, we may not be asking the right question. We know that the critical factor in ability to learn mathematics is intelligence and that "male" and "female" intelligence are very much the same, or certainly more alike than they are different. Thus, we are forced to overemphasize the small differences that show up. In trying to learn why some people do better at math than others, some researchers set the math achievement tests aside and look, instead, at the different dimensions of math ability: number sense, computation, spatial visualization, problem-solving skills, and mathematical reasoning. This might enable us to find out exactly where those differences in achievement are and greatly improve our chances of doing something about them.

Popular wisdom holds that females are better at computation and males at problem solving, females at "simple repetitive tasks" and males at restructuring complex ideas. However, since experts cannot even agree on what these categories are, still less on how to measure them, we have to be careful about accepting sex differences in "mathematical reasoning" or "analytic ability" as reported by the researchers in this field. It is fascinating to speculate that there are "innate capacities" to analyze or to reason mathematically, but these qualities have simply not been found.

What then do we know? As of 1978, are there any "facts" about male-female differences in mathematics performance that we can accept from the varied and not always consistent research findings? Possibly not, since the field is so very much in flux. But at least until recently, the "facts" were taken to be these:

Boys and girls may be born alike in math ability, but certain sex differences in performance emerge as early as such evidence can be gathered and remain through adulthood. They are:

1. Girls compute better than boys (elementary school and on).
2. Boys solve word problems better than girls (from age 13 on).
3. Boys take more math than girls (from age 16 on).
4. Girls learn to hate math sooner and possibly for different reasons.

One reason for the differences in performance, to be explored later in this chapter, is the amount of math learned and used at play. Another may be the difference in male-female maturation.* If girls do better than boys at all elementary school tasks, then they may compute better only because arithmetic is part of the elementary school curriculum. As boys and girls grow older, girls are under pressure to become less competitive academically. Thus, the falling off of girls' math performance from age 10 to 15 may be the result of this kind of scenario:

1. Each year math gets harder and requires more work and commitment.
2. Both boys and girls are pressured, beginning at age 10, not to excel in areas designated by society as outside their sex-role domain.
3. Girls now have a good excuse to avoid the painful struggle with math; boys don't.

Such a model may explain girls' lower achievement in math overall, but why should girls have difficulty in problem solving? In her 1964 review of the research on sex difference, Eleanor Maccoby also noted that girls are generally more conforming, more suggestible, and more dependent upon the opinion of others than boys (all learned, not innate behaviors).** Thus they may not be as willing to take risks or to think for themselves, two necessary behaviors for solving problems. Indeed, a test of third graders that cannot yet be cited found girls

*Girls are about two years ahead of boys on most indices of biological maturation throughout childhood.

**This is confirmed by Susan Auslander of the Wesleyan Math Clinic, whose "Analysis of Changing Attitudes toward Mathematics" (1978, unpublished) found that females place more value on outside opinion of success in mathematics than males.

nowhere near as willing to estimate, to make judgments about "possible right answers," and to work with systems they had never seen before. Their very success at doing the expected seems very much to interfere with their doing something new.

If readiness to do word problems, to take one example, is as much a function of readiness to take risks as it is of "reasoning ability," then there is more to mathematics performance than memory, computation, and reasoning. The differences between boys and girls—no matter how consistently they show up—cannot simply be attributed to differences in innate ability.

Still, if you were to ask the victims themselves, people who have trouble doing math, they would probably not agree; they would say that it has to do with the way they are "wired." They feel that they somehow lack something—one ability or several—that other people have. Although women want to believe they are not mentally inferior to men, many fear that in math they really are. Thus, we must consider seriously whether there is any biological basis for mathematical ability, not only because some researchers believe there is, but because some victims agree with them.

The Arguments from Biology

The search for some biological basis for math ability or disability is fraught with logical and experimental difficulties. Since not all math underachievers are women and not all women avoid mathematics, it is not very likely on the face of it that poor performance in math can result from some genetic or hormonal difference between the sexes. Moreover, no amount of speculation so far has unearthed a "mathematical competency" in some tangible, measurable substance in the body. Since masculinity cannot be injected into women to see whether it improves their mathematics, the theories that attribute such ability to genes or hormones must depend on circumstantial evidence for their proof. To explain the percent of Ph.D.'s in mathematics earned by women, we would have to conclude either that these women have different genes, hormones, and brain organization than the rest of us; or that certain positive experiences in their lives have largely undone the negative influence of being female; or both.

In her wide-ranging and penetrating review of the research on biological causes of mathematics ability, Julia Sherman captures the right historical perspective when she reminds us that as recently as a hundred years ago, it was assumed that because men are larger and heavier their brains must be larger and heavier, too.[9] When it was finally demonstrated that brain weight is irrelevant to brain power, the search for the causes of male intellectual superiority turned to secondary sexual differences: the incidence of gout in men, owing to excess uric acid, and the draining off by women's wombs of the "life forces" that fed the brain.

Sherman concludes from her review of the uric acid controversy (which still rages in some circles), that there is no likely connection between gout and genius, although excess uric acid, being the result of excess stress, could correlate with high activity and intelligence without causing it. As for the womb-brain controversy, historians and women students themselves have long since put this fantasy to rest. The fear was not so much that the brain would be starved by the womb as that the womb would be starved by the brain, and that on the eve of her manifest destiny, America would be populated by a generation of "puny men."

At the root of many of the assumptions about biology and intelligence is the undeniable fact that there have been fewer women "geniuses." The distribution of genius, however, is more a social than a biological phenomenon. An interesting aspect of the lives of geniuses is precisely their dependence on familial, social, and institutional supports. Without schools to accept them, men of wealth to commission their work, colleagues to talk to, and wives to do their domestic chores, they might have gone unrecognized—they might not even have been so smart. In a classic essay explaining why we have so few great women artists, Linda Nochlin Pommer tells us that women were not allowed to attend classes in art schools because of the presence of nude (female) models. Nor were they given apprenticeships or mentors; and even when they could put together the materials they needed to paint or sculpt, they were not allowed to exhibit their work in galleries or museums.[10]

Women in mathematics fared little better. Emmy Noether, who may be the only woman mathematician considered a genius, was honored (or perhaps mocked) during her lifetime by being called "Der Noether" ("Der" being the masculine form of "the"). Der Noether notwithstanding, the search for the genetic and hormonal origins of math ability goes on.

Genetically, the only difference between males and females (albeit a significant and pervasive one) is the presence of two chromosomes designated "X" in every female cell. Normal males have an "X-Y" combination. Since some kinds of mental retardation are associated with sex-chromosomal anomalies, a number of researchers have sought a link between specific abilities and the presence or absence of the second "X." But the link between genetics and mathematics is simply not supported by conclusive evidence.

Since intensified hormonal activity begins at adolescence and since, as we have noted, girls seem to lose interest in mathematics during adolescence, much more has been made of the unequal amounts of the sex-linked hormones, androgen and estrogen, in females and males. Estrogen is linked with "simple repetitive tasks" and androgen, with "complex restructuring tasks." The argument here is not only that such specific talents are biologically based (probably undemonstrable) but also that such talents are either-or; that one cannot be good at *both* repetitive and restructuring kinds of assignments.

Further, if the sex hormones were in any way responsible for our intellectual functioning, we should get dumber as we get older since our production of both kinds of sex hormones decreases with age.* But as far as we know, hormone production responds to mood, activity level, and a number of other external and environmental conditions as well as to age. Thus, even if one day we were to find a sure correlation between the amount of hormone present and the degree of mathematical competence, we would not know whether it was the mathematical competence that caused the hormone level to increase or the hormone level that gave

*Indeed, some people do claim that little original work is done by mathematicians once they reach age 30. But a counter explanation is that creative work is done not because of youth but because of "newness to the field." Mathematicians who originate ideas at 25, 20, and even 18 are benefiting not so much from hormonal vigor as from freshness of viewpoint and willingness to ask new questions. I am indebted to Stuart Gilmore, historian of science, for this idea.

us the mathematical competence.

All this criticism of the biological arguments does not imply that what women do with their bodies has no effect on their mathematical skills. As we will see, toys, games, sports, training in certain cognitive areas, and exercise and experience may be the intervening variables we have previously mistaken for biological cause. But first we must look a little more closely at attitude.

Sex Roles and Mathematics Competence

The frequency with which girls tend to lose interest in math just at puberty (junior high school) suggests that puberty might in some sense cause girls to fall behind in math. Several explanations come to mind: the influence of hormones, more intensified sex-role socialization, or some extracurricular learning experience boys have at that age that girls do not have. Having set aside the argument that hormones operate by themselves, let us consider the other issues. Here we enter the world of attitudes, as formed by experience and expectation.

One group of seventh graders in a private school in New England gave a clue to what children themselves think about this. When visitors to their math class asked why girls do as well as boys in math until sixth grade but after sixth grade boys do better, the girls responded: "Oh, that's easy. After sixth grade, we have to do real math." The reason why "real math" should be considered accessible to boys and not to girls cannot be found in biology, but only in the ideology of sex differences.

Parents, peers, and teachers forgive a girl when she does badly in math at school, encouraging her to do well in other subjects instead. " 'There, there,' my mother used to say when I failed at math," one woman remembers. "But I got a talking-to when I did badly in French." "Mother couldn't figure out a 15 percent tip and Daddy seemed to love her more for her incompetence," remembers another. Lynn Fox, who has worked intensively in a program for mathematically gifted teenagers who are brought to the campus of Johns Hopkins University for special instruction, finds it difficult to recruit girls and to keep them in her program. Their parents sometimes prevent them from participating altogether for fear it will make their daughters too different, and the girls themselves often find it difficult to continue with mathematics, she reports, because they experience social ostracism. The math anxious girl we met in Chapter Two, who would have lost her social life if she had asked an interesting question in math class, was anticipating just that.

Where do these attitudes come from?

A study of the images of males and females in children's textbooks by sociologist Lenore Weitzman of the University of California at Davis, provides one clue to why math is associated with men and boys in the minds of little children.*[11] "Two out of every three pictures in the math books surveyed were of males, and the examples given of females doing math were insulting and designed to reinforce the worst of the stereotypes," she reports (Fig. III-1 and III-2).

*The study, done in 1975, reviewed textbooks in current use in the California school system. Eight thousand pictures were sorted and coded by sex, race, and age of the persons in the picture; also by what they were doing. It is available as a slide show from the National Education Association in Washington, D.C. The pictures included here are composites, typical, but not taken from any one book.

FIG. III-1

Weitzman comments: "It seems ironic that housewives who use so much math in balancing their accounts and in managing household budgets are shown as baffled by simple addition."

"Another feature of the mathematics textbooks," says Weitzman, "is the frequent use of sex as a category for dividing people, especially for explaining set theory" (Fig. III-3).

"When sex is used as a category, girls are told that they can be classified as different," Weitzman believes,

FIG. III-2

girls

people who like to cook

FIG. III-3

"as typically emotional or domestic . . . There is also strong sex typing in the examples used and in the math problems" (Fig. III-4 and III-5).

FIG. III-4 FIG. III-5

"We found math problems," Weitzman writes, "in which girls were paid less than boys for the same work. It would be hard to imagine a textbook publisher allowing this example if a black boy were being paid less than a white boy. Yet it seems legitimate to underpay girls." (Fig. III-6).

In another survey of math textbooks published in 1969, not one picture of a girl was found and the arithmetic problems used as examples in the book showed adult women having to ask even their children for help with math, or avoiding the task entirely by saying, "Wait until your father comes home."*

Adults remember their junior high school experi-

*As a matter of interest, it is not necessary to use boys or girls or mommies or daddies in arithmetic problems. Holt, Rinehart and Winston and Addison-Wesley publish texts with problems like these: "Sold six fish; bought two more. How many now?"

FIG. III-6

ences in math as full of clues that math was a male domain. No so long ago, one junior high school regional math competition offered a tie clasp for first prize. A math teacher in another school, commenting unfavorably on the performance at the blackboard of a male student, said to him, "You think like a girl." If poor math thinkers think like girls, who are good math thinkers supposed to be?

The association of masculinity with mathematics sometimes extends from the discipline to those who practice it. Students questioned about characteristics they associate with a mathematician (as contrasted with a "writer") selected terms like rational, cautious, wise, and responsible. The writer, on the other hand, in addition to being seen as individualistic and independent, was also described as warm, interested in people, and altogether more compatible with a feminine ideal.[12]

As a result of this psychological conditioning, or what Lynn Osen calls the "feminine math-tique," a young woman may consider math and math-related fields inimical to femininity. In an interesting study of West German teenagers, Erika Schildkamp-Kuendiger found that girls who identified themselves with the feminine ideal underachieved in mathematics, that is, did less well than was expected of them on the basis of general intelligence and performance in other subjects.[13]

Thus, the problem of girls' poor performance in math is by no means limited to America, though it is also *not* found in a few countries. "Men in the Soviet Union," writes Norton Dodge, "are so accustomed to women's participating in all fields of study, that the performance of girls is comparable to that of boys in mathematics and physics."[14]

Is there anything about sex roles and mathematics that is not culturally relative?

Interestingly, research shows that intelligence in general (and possibly even mathematical intelligence would fit this model) correlates not with extreme masculinity or femininity but with cross-sex identification. Boys and girls who pursue some of the interests and behaviors of the opposite sex score higher on general intelligence tests and tests of creativity than children who are exclusively masculine or feminine.[15] Girls who resist the pressure to become ladylike and instead develop aggressiveness, independence, self-sufficiency, and tough-mindedness score higher on these tests than more passive, "feminine" girls. And similarly, boys who are "sensitive" score higher than more typical, aggressive boys.

Average boys may feel the need to misbehave in school because they are getting mixed messages from their parents: be naughty on the playing field but quiet and controlled in school. Yet girls are expected to be consistently docile. Another possibility is that children of both sexes who are more intelligent to begin with may find society's rules cumbersome and choose to ignore them. That is, either intelligence breeds behavior that is not typical of the child's sex, or unconventional experience stimulates intelligence, or both.

There is far less data about personality characteristics related to mathematical performance than about personality and intelligence in general. But one recent study compared three groups of women college students, one group higher in verbal abilities than in math, the second group higher in math than in verbal, and the third group about equally competent in both. The study found that the women higher in math ability responded positively to a cluster of attributes considered masculine, such as "logical," "persistent," and "intellectual." But this group also scored high on positively valued feminine attributes, such as "warm," "generous," and so on. The researcher concluded that these women were not rejecting femininity itself, but such low-valued feminine characteristics as dependence and passivity. The women seemed to have a healthy orientation toward the best of both the male and female worlds.[16]

Such a conclusion is an important modification of past studies that found women who do well in math to be "masculine." One such study, linking problem-solving ability in math with a masculine self-image, even went so far as to conclude that nonmathematical men had an image problem. Another early analysis of the autobiographies of women in mathematics concluded that these women either lacked a "typical feminine identification" or were "conflicted" over their female role. Far better and more recent studies of women mathematicians do not explain their success by their masculine or feminine "nature," but find that these women enjoy some real, tangible advantages, among them strong family support.

If sex-role socialization is what one is taught about oneself by others, then we may call what one learns about oneself by oneself "experience." And we must look into the impact of *experience* on learning math.

One appealing theory sounds almost too simple. It is that people who do well in mathematics from the beginning and people who have trouble with it have altogether different experiences in learning math. These differences are not necessarily innate or cognitive or even, at the outset, differences in attitude or in appreciation for math. Rather, they are differences in how people cope with uncertainty, whether they can tolerate a certain amount of floundering, whether they are willing to take risks, what happens to their concentration when an approach fails, and how they feel about failure. These attitudes could be the result of the kinds of risks and failures they remember from early experience with mathematics, because our expectations of ourselves are shaped not simply by what others say but by what we think we can and cannot do.

Street Mathematics: Things, Motion, and Scores

If a ballplayer is batting .233 going into a game and gets three hits in four times at bat (which means he has batted .750 for the day), someone watching the game might assume that the day's performance will make a terrific improvement in his batting average. But it turns out that the three-for-four day only raises the .233

to .252. Disappointing, but a very good personal lesson in fractions, ratios, and percents.

Scores, performances like this one, lengths, speeds of sprints or downhill slaloms are expressed in numbers, in ratios, and in other comparisons. The attention given to such matters surely contributes to a boy's familiarity with simple arithmetic functions, and must convince him, at least on some subliminal level, of the utility of mathematics. This does not imply that every boy who handles runs batted in and batting averages well during the game on Sunday will see the application of these procedures to his Monday morning school assignment. But handling figures as people do in sports probably lays the groundwork for using figures later on.*

Not all the skills necessary for mathematics are learned in school. Measuring, computing, and manipulating objects that have dimensions and dynamic properties of their own are part of everyday life for some children. Other children who miss these experiences may not be well primed for math in school.

Feminists have complained for a long time that playing with dolls is one way to convince impressionable little girls that they may only be mothers or housewives, or, in emulation of the Barbie doll, pinup girls when they grow up. But doll playing may have even more serious consequences. Have you ever watched a little girl play with a doll? Most of the time she is talking and not doing, and even when she is doing (dressing, undressing, packing the doll away) she is not learning very much about the world. Imagine her taking a Barbie doll apart to study its talking mechanism. That's not

*An early analysis of the possibility that performance differences between the sexes could arise from differences in games and activities is in Julia Sherman, "Problems of Sex Differences in Space Perception and Aspects of Intellectual Functioning," *Psychological Review,* 1964, *74,* pp. 290–299

the sort of thing she is encouraged to do. Do girls find out about gravity and distance and shapes and sizes playing with dolls? Probably not!

A college text written for inadequately prepared science students begins with a series of supposedly simple problems dealing with marbles, cylinders, poles made of different substances, levels, balances, and an inclined plane. Even the least talented male science student will probably be able to see these items as objects, each having a particular shape, size, and style of movement. He has balanced himself or some other object on a teeter-totter; he has watched marbles spin and even fly. He has probably tried to fit one pole of a certain diameter inside another, or used a stick to pull up another stick, learning leverage. Those trucks little boys clamor for and get are moving objects. Things in little boys' lives drop and spin and collide and even explode sometimes.

The more curious boy will have taken apart a number of household and play objects by the time he is ten; if his parents are lucky, he may even have put them back together again. In all this he is learning things that will be useful in physics and math. Taking out parts that have to go back in requires some examination of form. Building something that stays up or at least stays put for some time involves working with structure. Perhaps the absence of things that move in little girls' childhoods (especially if they are urban little girls) quite as much as the presence of dolls makes the quantities and relationships of math so alien to them.

In sports played, as well as sports watched, boys learn more math-related concepts than girls do. Getting to first base on a not very well hit grounder is a lesson in time, speed, and distance. Intercepting a football in the air requires some rapid intuitive eye calculations based on the ball's direction, speed, and trajectory. Since physics is partly concerned with velocities, trajectories, and collisions of objects, much of the math taught to prepare a student for physics deals with relationships and formulas that can be used to express motion and acceleration.* A young woman who has not closely observed objects travel and collide cannot appreciate the power of mathematics.

Unfamiliarity with things may also cause a girl to distrust her environment. Since the movement of objects seems not only irregular but capricious, watching things move may not seem to her to be as reliable a way to learn about the world as following the lesson in a book.** A wilderness canoe instructor confirms this as he describes a woman learning to canoe:

> When I start my preliminary instruction, she hangs on to my words, watching me intently. When she gets into the canoe, she mimics exactly what I have done, even if it is inappropriate . . . She wants to know how to put the paddle in the water, how hard to pull on it, when to start pulling, where to hold it . . . She makes the operation into a ritual like a dance, becoming increasingly tense (and frustrated) as time goes on . . .
>
> At some point I say to her, "Now look, you are trying to get your feedback from the wrong place. You keep watching me and you should be watching the boat. The boat will tell you what is right and what is wrong. When you do it right the boat does what you want it to.
>
> "I can tell you what to look for, but I can't tell you how it

*In learning physics, unlike math, however, intuitive notions have to be unlearned: for example, heavier objects will not always fall faster. So, to some extent, real-life experience may be counterproductive.

**Even the arithmetic games that girls like to play will hardly teach them about the world of natural physical events. Monopoly and playing store provide practice in arithmetic fundamentals, but nothing that might suggest some of the more complex phenomena in her environment that mathematics can explain.

feels. As long as you keep watching me you will get nowhere. Now forget about being instructed. Just go out in the canoe and play around with it and find out what it does."[17]

The point is that what we get out of an experience, even a good one, may depend on what we have done and learned before. One thing is what you know you *can* do. Another is what you think you *should* do; and the combination of limited physical experience and negative attitudes toward math may be the principal contributor to females' poorer performance in mathematics.

Conclusion

After surveying the summaries of research in this area and interviewing people who claim to be incompetent at mathematics, I have reached a conclusion. Apart from general intelligence, which is probably equally distributed among males and females, the most important elements in predicting success at learning math are motivation, temperament, attitude, and interest. These are at least as salient as genes and hormones (about which we really know very little in relation to math), "innate reasoning ability" (about which there is much difference of opinion), or number sense. This does not, however, mean that there are no sex differences at all.

What is ironic (and unexpected) is that as far as I can judge sex differences seem to be lodged in *acquired skills;* not in computation, visualization, and reasoning *per se,* but in ability to take a math problem apart, in willingness to tolerate certain kinds of ambiguity, and in careful attention to mathematical detail. Such temperamental characteristics as persistence and willingness to take risks may be as important in doing math as pure memory or logic. And attitude and self-image, particularly during adolescence when the pressures to conform are at their greatest, may be even more important than temperament. Negative attitudes, as we all know from personal experience, can powerfully inhibit intellect and curiosity and can keep us from learning what is well within our power to understand.

An Afterword: Math Anxiety or Math Avoidance—How Can We Tell the Difference?

There is ample evidence that avoidance of mathematics is disproportionate among girls and women, beginning with the eleventh grade and extending through every stage of their educational and professional development. A recent study of women Ph.D. candidates in political science found that of all the factors that might have influenced these women's choice of graduate school (including academic standards, prestige, percentage of women on the faculty, financial aid availability, etc.), only one factor in every case predicted the school the women selected: whether there was a mathematics or statistics requirement.[18] Was this math avoidance rational or does it indicate severe math anxiety?

It may be, as some argue today, that math anxiety is only forgetfulness, unfamiliarity, and awkwardness in returning to a subject one has not studied for a very long time, that it is not so much a *cause* of math avoidance as an *effect.* Leaving that question for the moment, are there dangers in emphasizing "anxiety" over "avoidance?" Does such a term, particularly when applied to women, imply that women are more impressionable and weaker in spirit than men? Several feminists criticize the anxiety model, pointing out that since the causes of math anxiety lie in "political and social forces that oppress women" and are not wholly psychological and educational in origin, the goal of remediation should not be "the curing of an individual case but the elimination of the conditions that foster the disease."[19]

The identification of mathematics anxiety as a problem for women could become two-edged. Focusing on one more female "disability" may feed the prejudices that already abound in the real world about women and math, women and science, and women and machines. We also have to consider the needs of women who are very competent in math and have a hard time proving this to their colleagues. Finally, we have to contemplate the possibility that attention given to this issue might expose women to exploitation by "math anxiety experts."

Mindful of all these objections, I still argue that excessive anxiety inhibits women more than it does men. If you ask any female how she feels about mathematics, you may find out how she feels about many other gender-related aspects of her life as well. Very bright girls who excel at almost everything in school feel quite comfortable failing at math, not simply because their parents allow it and their peers accept it, but because it provides a solution to the conflicts their brightness creates for them. Rebellious adolescent girls, on the other hand, may actually force themselves to like and do well at math as a way of holding their femininity at bay for a while.

Trying to help one young woman graduate student overcome her intense hostility to math, Stanley Kogelman, co-founder of Mind over Math, heard her say that it was the "logic" and "discipline" of mathematics that she disliked most. Probing to find out where those feelings about "logic" and "discipline" come from, Kogelman concluded that the woman was really disturbed by the fear that she would enjoy the rigorous part of her *own* mind. Mathematics was incidental in her struggle. She was actually in conflict over her own identity.[20]

Do men suffer from math anxiety, and does it intrude as much on their lives as it does for women? Until we have more satisfactory measures of math anxiety and have more math autobiographies from men, we will not have much to say about this. But one recent dissertation study does suggest that although men have math anxiety too, it doesn't trouble them quite as much. The study was done with 655 Ohio State University undergraduates enrolled in a precalculus course. The researcher tested their math anxiety (as best she could with a paper-and-pencil questionnaire) and then compared their anxiety ratings with their final grades. She found, interestingly, that the men's math anxiety scores did *not* correlate with their final course grades nearly as much as the women's anxiety scores correlated with theirs.[21] Perhaps the men went to greater lengths to hide their anxiety even from the researcher. Or perhaps, as the researcher concludes, math anxiety is harder for women to overcome.

My hunch is that the researcher is right. Men have math anxiety too, but it disables women more.[22]

References

[1]Gerard Piel, "Taking Down the Hurdles," unpublished speech, Washington, D.C., American Academy of Arts and Sciences, February, 1976.
[2]Lewis Aiken, "Review of the Literature on Attitudes Toward Mathematics," *Review of Educational Research, 46,* no. 2 (Spring, 1976), p. 293–311.
Lorelei Brush, "A Path Analytic Explanation of Lower Participation Rates of Women in Physical Science Courses," unpublished paper, 1976.
———, "Summary of Research on Students' Avoidance of Mathematics or the Humanities," unpublished report to school personnel, undated.
———, "A Validation Study of the Mathematics Anxiety Rating Scale," *Educational and Psychological Measurement,* in press.
John Ernest, "Mathematics and Sex," *American Mathematical Monthly, 83,* no. 8 (October 1976), pp. 595–614.
———, *Is Mathematics a Sexist Discipline?* Summary of a paper presented at the annual meeting of the American Association for the Advancement of Science, February 18, 1976.
Elizabeth Fennema, "Influences of Selected Cognitive, Affective, and Educational Variables on Sex-Related Differences in Mathematics and Learning and Studying," prepared with the assistance of Mary Ann Konsin for the National Institute of Education, under Grant P-76-0274, October, 1976.
——— and Julia Sherman, "Sex-Related Differences in Mathematics Achievement, Spatial Visualization and Affective Factors," *American Educational Research Journal, 14,* no. 1 (Winter, 1977), pp. 5–71.
Lynn Fox, "The Effects of Sex Role Socialization on Mathematics Participation and Achievement," paper prepared for the National Institute of Education, Contract FN-17-400-76-0114, December, 1976.
Michael Nelson, "Mathematical Ability, Inability, and Disability," unpublished master's thesis, Alfred P. Sloan School of Management, MIT, 1977.
Lynn Osen, "The Feminine Math-tique," Pittsburgh, K.N.O.W., 1971.
Julia Sherman, "Effects of Biological Factors on Sex-Related Differences in Mathematics Achievement," prepared for the National Institute of Education, Contract 400-76-0113, Fall, 1976.
[3]For the most comprehensive review of research on sex differences, see the two books by Eleanor Maccoby (the second with Carol Nagy Jacklin):
E.E. Maccoby, *Development of Sex Differences,* Stanford, Stanford University Press, 1966.
——— and C. N. Jacklin, *Psychology of Sex Differences,* Stanford, Stanford University Press, 1974.
[4]Ernest, "Mathematics and Sex," *op. cit.,* p. 597.
[5]Fennema, *op. cit.,* p. 87.
[6]The National Assessment of Performance and Participation of Women in Mathematics, is a two-year study of a national sample of 5,000 13- and 17-year-old students, male and female, beginning in the fall of 1978. It will be conducted by the National Assessment of Educational Progress,

Office of the Assistant Secretary, U.S. Department of Health, Education, and Welfare.
[7]Brush, "Validation Study," p. 5.
[8]Fennema, *op. cit.,* p. 7.
[9]Sherman, *op. cit.,* p. 137.
[10]Linda Nochlin Pommer, "Why Are There No Great Women Artists," *Art News,* January, 1971.
[11]Lenore Weitzman and Diane Rizzo, "Images of Males and Females in Elementary School Textbooks," Washington, NEA, 1975.
[12]Brush, "Validation Study," p. 7. See also Judith E. Jacobs, "A Comparison of the Relations of Acceptance of Sex-Role Stereotyping and Achievement and Attitudes towards Math of 7th and 11th graders in a Suburban Metropolitan New York Community," unpublished dissertation, New York University, 1973.
[13]Erika Schildkamp-Kuendiger, *Die Frauenrolle und die Mathematikleistung,* Dusseldorf, Schwann, 1974.
[14]*Norton* Dodge as quoted in Osen, *op. cit. passim.*
[15]E.E. Maccoby, *Development of Sex Differences,* 1966, *op. cit.,* p. 43.
[16]Nancy Potter, "Mathematical and Verbal Ability Patterns in Women," unpublished dissertation, University of Missouri–Columbia, 1974, *passim.*
[17]John D. McRuer, canoeist and instructor for the Algonquin Waterways Wilderness Tours, Toronto, Canada. Letter to author, 1976.
[18]Barbara Merrill, "Affirmative Action and Women in Academia: Factors Determining the Number of Women Enrolled in Political Science Ph.D. Programs," Unpublished dissertation, Ohio State University, Spring, 1976, *passim.*
[19]Patti Hague, "Technology Assessment from a Feminist Perspective: Math Anxiety Programs," unpublished article, 1976, *passim.* See also Judith Jacobs, "Women and Mathematics: Must They Be at Odds," Pi Lambda Theta *Newsletter,* September, 1977.
[20]Stanley Kogelman, "Debilitating Mathematics Anxiety: Its Dynamic and Etiology," Unpublished Master's Thesis, Smith College School of Social Work, 1975.
[21]Nancy E. Betz, "Math Anxiety: What Is It?" Paper delivered at the American Psychological Association Convention, San Francisco, 1977, *passim.*
[22]See also the author's, "Math Anxiety: Why Is a Smart Girl Like You Still Counting on Your Fingers?" *Ms.,* September, 1976, p. 56; and with Bonnie Donady, "Counselling the Math Anxious," *Journal of the American Women Deans, Counsellors and Administrators,* Fall, 1977, p. 13; and with Bonnie Donady, "Math Anxiety," *Teacher,* November, 1977, p. 71.
Knowles Dougherty, "Sex Differences in Word Problem Solving as a Function of Age," in Richard Purnell, Ed. *Adolescents and the American High School,* New York, Holt, Rinehart and Winston, 1970, pp. 96 ff.

4

Right- and Wrongheadedness: Is There a Nonmathematical Mind?

Spatial Visualization

As we have seen, not all women experience math anxiety and not all people who fear mathematics are women. This is essential to the proper understanding of the preceding chapter. There is, however, one knotty facet of intelligence that prevents researchers from abandoning altogether the study of male-female differences in intellectual activities. Tests of the ability to understand and manipulate drawings of two-dimensional and three-dimensional figures generally show males to be more skilled in this area than females. This skill is called spatial visualization. Even among very intelligent girls and women, the capacity to visualize shapes moving through space is significantly less well developed than in comparably intelligent males. When large groups are compared, males on the average do better than females at finding embedded figures hidden in complex drawings, solving geometric problems, "cutting" cubes, learning mazes, and reading maps. Researchers have even found that telling left from right varies to some extent by sex. Try out your own spatial visualization on the tests beginning with Fig. IV-1.

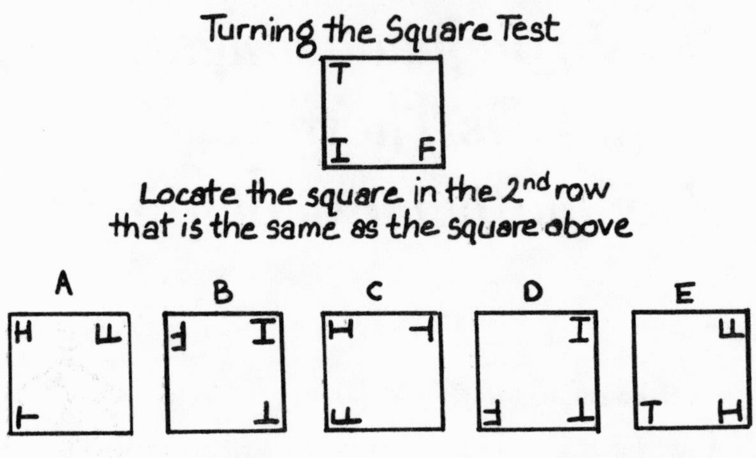

FIG. IV-1

The first example in two dimensions requires a simple rotation. The goal is to find the image in the collection A, B, C, D, E that is exactly the same as the one in the example. In each example the image has been rotated, but the letters must be in the same relation for a correct answer. If you have to "translate" the operation into terms like these:

The stem of the *T* faces the right-hand vertical of the *H* and the *H* is in the very next corner to the *T;* therefore, the only items among the group of possible answers where these two conditions are met are B, C, and D. Since there are three possibilities, I need to look also at the *F.* Okay, the long side of the *F* faces the brace of the *H.* This eliminates C among the answers and also D. The answer must be B. (And it is.)

then you are tackling the problem verbally and not spatially. To do it spatially you should be able just to etch the image of the one into your mind and find the right answer by mentally rotating the image. Try Fig. IV-2.

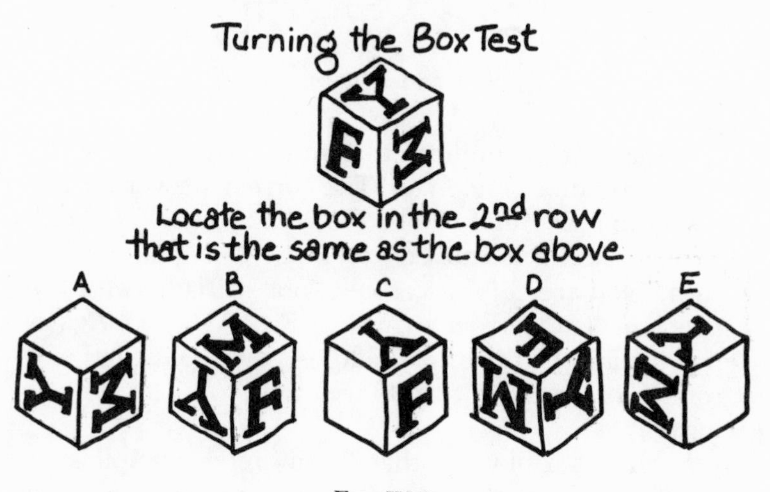

FIG. IV-2

Here the item is in three dimensions, but the rotation operation is similar. Again, if you have to talk out the relationship between the stem of the *Y* and the side of the *M* and note that the three invisible sides of the cube face the long stem of the *F,* the right side of the *Y,* and the top side of the *M,* you are verbalizing the test. The correct answer is not A because the stem of the *Y* meets the bottom side of the *M.* B is incorrect because when the *M* is on top the *Y* would be invisible. C is not correct either, because the *Y* and the *F* are in the wrong relation. D is right.

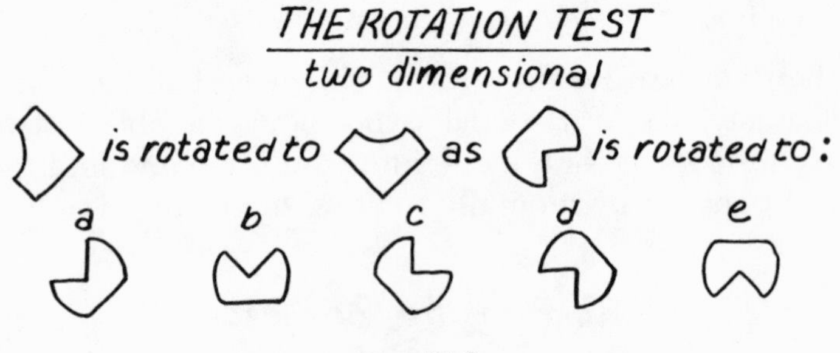

FIG. IV-3

Try a more complicated item involving flipping as well as rotation (Fig. IV-3). The correct answer for this test item is D. But if you had to say to yourself, as I did, "The example is moved clockwise one-quarter of a turn," you are coping verbally, not spatially, with the problem. The correct answer in Fig. IV-4 is D. Again, if you said, "The example is flipped over to the left and rotated one-quarter turn to the right," you are "cheating."

I think it is obvious that in any more complicated items than these, the rotations, flips, and turns would be virtually impossible to talk through. Hence, at some point in such a spatial skills test, either we would see the right answer or we would have to stop.

Try not talking through the next two examples (Figs. IV-5 and IV-6).

There are very few measurable sex differences in intelligence. Therefore much has been made of this one factor, spatial intuition, and particularly of its possible bearing on analytic thinking. Since there is even speculation that poor spatial visualization alone can account

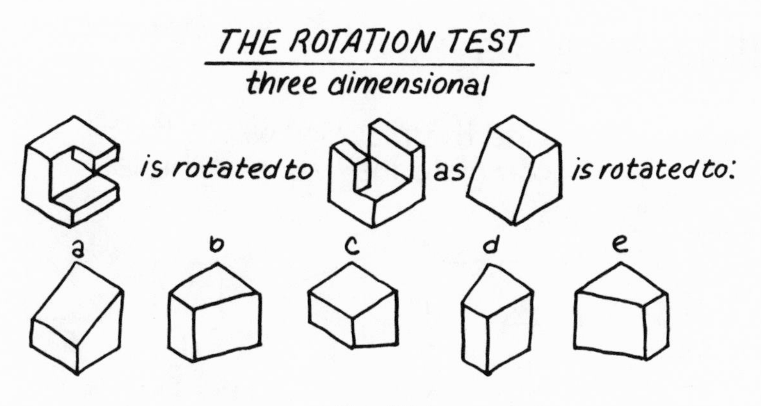

FIG. IV-4

for difficulty in learning advanced mathematics, anyone trying to understand the greater incidence of math avoidance and math anxiety in women must consider the possibility that their poorer spatial skills are the critical factor. But is there sufficient evidence that all or most women do less well than men? And can we definitely link the ability to see spatial relations and math? These are the critical issues.

During World War II, women were not entirely trusted as trained pilots. Although some women did fly freight missions (and Amelia Earhart had previously won trophies for her even more daring flights), the reason given for not training more women as pilots was that they were not as good as men at spatial visualization. They tended not to be able to right themselves while blindfolded and strapped to a tilted chair in a tilted room. Many other occupations have also used some form of spatial skills testing on an entrance examination and, perhaps unintentionally, as a way to exclude women. Even now, most dental schools test spa-

Finish the Box Test

Find the piece or pieces that complete the figure on the left

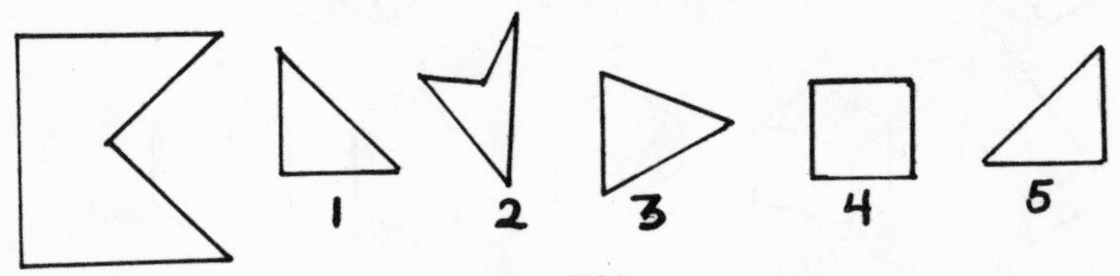

FIG. IV-5

Fold the Box Test

Find the folded box that represents the unfolded box on the le

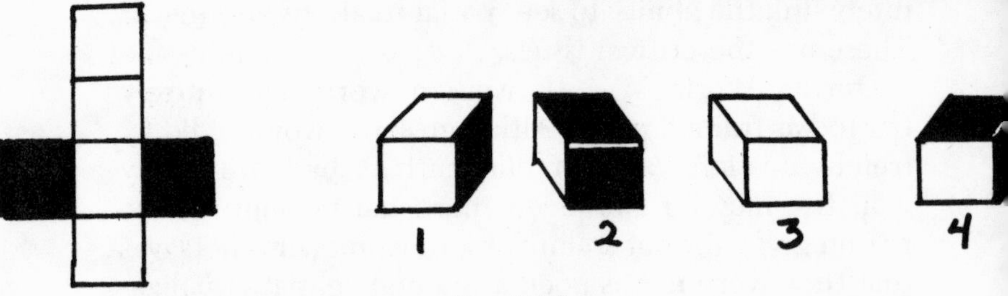

FIG. IV-6

tial visualization ability not by having applicants carve a plaster cast, as was once required, but with a paper and pencil test including items similar to the ones we have just examined.

This is one misuse of the spatial ability test, since

there is no firm evidence that persons who do poorly on the test do not become competent dentists. Another misuse is in the assumed connection between analytic ability—a much more widely appreciated talent than mere spatial ability—and spatial visualization. A frequently used measure of analytic ability is the "embedded figure test."

In this test (Fig IV-7), the goal is to find among an assortment of items the one in which the shape given at the outset is embedded. Since the shape is mixed with an array of confusing lines, what is supposedly being tested is the person's ability to sort out the item from its context, the essential characteristic from its confusing environment. The problem with this test as a measure of analytic ability, as Julia Sherman has pointed out, is that the characteristic is not an idea or a theme or an inconsistency—the sort of things a good analyst must sort out from confusing data—but rather a *shape.* Sherman argues powerfully that the embedded figure test should not be used because it measures two "abilities," spatial visualization and analytic capability, not one. Thus it is unfair to women, who are not as good as men at spatial visualization. On other tests of analytic ability where the item to be located is a key sentence in a paragraph or a number out of place in a series, women do as well as men.

But if sex differences in analytic ability do not exist (as most of us hope and believe), there is still some evidence that spatial skills seem less well developed in women (on the average) than in men. Where does spatial intuition come from? Is it lodged in the brain, as some researchers think? Or is it learned, much as boys learn to manipulate objects and scores through their play, their physical environment, and their expectations of themselves?

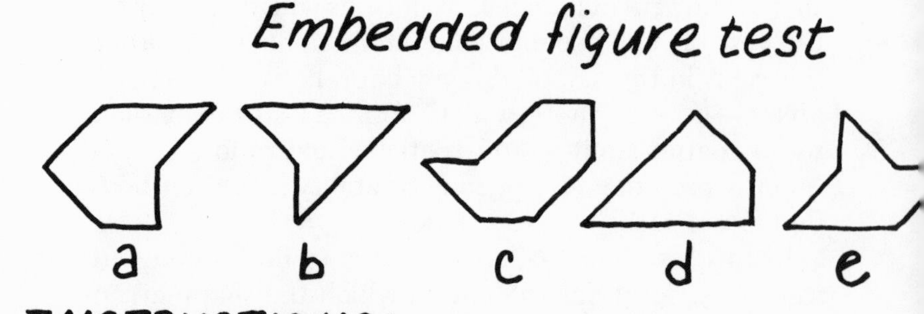

INSTRUCTIONS:
find the figure among those outlinec
above in the box or boxes below

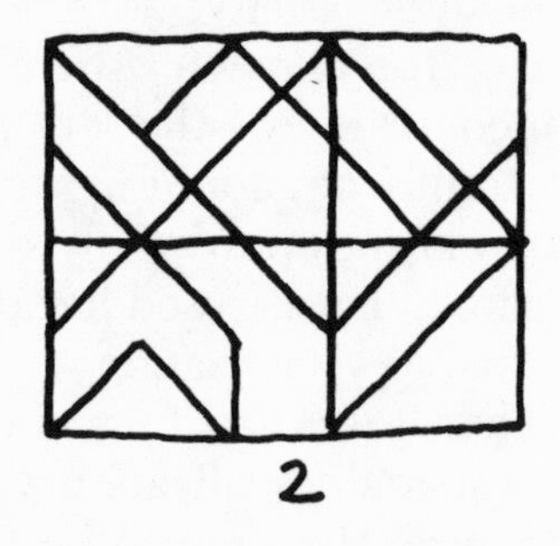

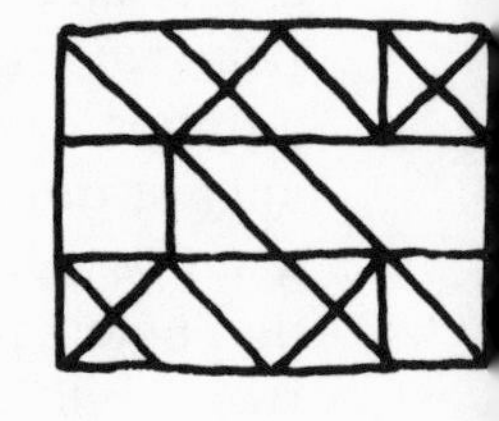

FIG. IV-7

The Theory of the "Two Hemispheres"

Whether spatial visualization can be learned and improved and whether this skill is related to mathematics are issues that have been debated for a very long time. But one area of research, on the structure and organization of the brain itself, is fairly new. Researchers now think that they can locate the seat of spatial visualization in the brain and can explain why certain women and men do not have well developed spatial faculties.[1]

During the past few decades several independent lines of research have been generating a new model of the brain. In this model, the brain consists of two relatively independent information processors—a right hemisphere and a left hemisphere—that function in very different ways. Joseph Bogen, a neurosurgeon on the West Coast, has been working with several kinds of brain damaged people: stroke victims who are paralyzed on one side of their bodies and epileptics whose brain hemispheres have been split surgically to prevent the seizures from spreading from one part of the brain to the other. By giving these patients very specific intellectual tasks to perform, Bogen has established the duality of the right and left hemispheres. The right hemisphere of the brain, which dominates the left side of the body, seems to contribute the abilities to perceive shapes, to remember musical phrases, to grasp things (though not necessarily ideas) as wholes, and to recognize faces. The left hemisphere, which dominates the right side of the body, specializes in speech and other linear or sequential tasks. Letters, Bogen points out, have to follow a certain order to make words, words must follow a certain order to make sentences, and the ability to grasp these sequences seems to be lodged in the left hemisphere. A person whose left hemisphere has been damaged by a stroke cannot speak or move the right side of his body. Yet he may well be able to sing and remember lyrics. Conversely, if the right hemisphere is malfunctioning, the patient will be able to read and write but may not recognize faces or remember geographical locations.[2]

Nevertheless, people who have had a hemisphere disconnected either through brain damage or surgery can do both right-hemisphere and left-hemisphere tasks. But they may not be able to utilize with one side of the brain information that went into the other. One startling example is a patient with a dysfunctional left hemisphere trying unsuccessfully to reproduce with his right hand a picture he has just looked at. Because the picture went into his right hemisphere (where all pictures go), he can only draw it with his left hand, which is directed by the right side of the brain. In a dramatic slide presentation, Bogen demonstrates that after a few minutes' struggle with his right hand (and he is naturally right-handed), the patient easily reproduces the picture with his left hand.

Bogen reports that the particular disabilities of the severed brain were not anticipated when this kind of surgery was introduced. Neurosurgeons expected to find a wall of tissue between the two hemispheres of the brain. Instead, they found what Bogen describes as a billion telephone lines, each linking some special function in one half of the brain to a function in the other. From this physical evidence, doctors came to believe that cross-brain communication is much more important in processing and using information than had previously been thought.

At about the time of this medical research, Robert Ornstein, a psychologist, was studying the brain functioning of normal individuals. Using a mechanism that measures the alpha waves being emitted by each half of the brain, he attached plates that could register these electrical charges to different points on his subjects' heads. Knowing that alpha rhythms increase when the brain is at rest, he noted that when the charge monitor went up it meant that the part of the brain to which it was attached was not being used. Monitoring the charges while giving his subjects particular tasks to do, Ornstein was able to associate those activities with one or the other brain hemisphere. His research confirmed that verbal and numerical tasks apparently make heavy (if not exclusive) use of the left hemisphere, while the right hemisphere rests, and that reproduction of designs, grasping whole pictures, and recognizing faces occur mainly in the right hemisphere.[3]

Bogen and Ornstein are more cautious than the popularizers of the "split brain" theory. They point out that hemisphericity, as they define it, should not be misinterpreted to mean that certain skills are lodged in one hemisphere and other skills in the other (although one can readily see how this conclusion could be drawn from the kinds of experiments Ornstein has done). Just because we know what we *cannot* do after the right hemisphere has been damaged does not mean that we know what the right hemisphere can do *by itself.*

Thinking, as we know it, embraces acquisition of knowledge, memory, reflection, abstraction, and creativity. It relies on the healthy development of both hemispheres and, above all, on the unimpeded interaction of the two. Reading difficulties, according to this model of the brain, may result from weakness in the tissue between the two hemispheres rather than a poorly functioning left hemisphere. Such a weakness might prevent an otherwise normally functioning child from connecting the *picture* of a word, perceived through the right hemisphere, with the *sound* of that word that has been recorded in his auditory memory in the left hemisphere. The implications of this research for understanding and treating brain abnormalities are enormous. But what, if anything, does split-brain research contribute to our understanding of mathematics learning and possible sex differences in learning spatial skills?

Brain Lateralization

So far nothing in split-brain research would especially account for male-female differences in spatial visualization. No more females than males are left-handed (a condition which in some but not all cases causes the hemispheres to reverse their functions). Most females, like most males, have speech lodged in the left hemisphere and the ability to perceive block designs lodged in the right.

But one sex difference has been noted in the research on brain specialization, and that is in the *timing* of its development in each individual. We are born, it seems, with the potential to do all the known kinds of mental functioning in both halves of the brain. Children who experience a blow to the left side of their head severe enough to cause left hemisphere brain damage often learn to speak using the right hemisphere, so long as the damage occurs before the age of seven or eight. But as we continue to grow the adaptability of the brain's hemispheres comes to an end. Older children who receive severe brain damage to the left hemisphere may never develop speech or learn to read because by then their right hemisphere (the undamaged one) has lost its potential for verbal tasks. There is now some speculation that, mysteriously, this final specialization or brain lateralization takes place earlier in girls than in boys. This, it is argued, accounts for girls' earlier verbal aptitude; but the fixing of the hemispheres also causes a stagnation of their spatial capability at an earlier, perhaps too premature, stage of development.

It is not altogether clear whether it is the *removal* of the spatial capability from the left hemisphere that matters most or the fact that the right hemisphere (in most people, the spatial hemisphere) *stops growing.* But in either case the assertion that girls lateralize earlier and therefore miss a crucial stage of spatial visualization development tends to buttress the arguments that a biological difference accounts for math avoidance and anxiety in females. Therefore, the argument must be examined rather carefully.

Let's review the theory. Both boys and girls have bilateral capacity for spatial visualization at birth. Maximal verbal and spatial functioning are supposedly attained when the hemispheres are lateralized or specialized. Girls develop this specialization earlier, but for most girls the specialization hastens their ability to read but removes spatial skills prematurely from the left hemisphere and leaves them only reduced spatial visualization in the right. In summary, although specialization is good, it is preferable to keep the capability for spatial visualization in both hemispheres as long as possible. Since boys are said to enjoy this advantage over girls, boys do better at spatial skills. From the point of view of this research, girls are overspecialized in left-hemisphere functioning and do spatial visualization best, if at all, by talking it through. As we noticed, to some extent a spatial test can be done this way, but it is far slower and not really what is expected.

The entire issue of brain lateralization is controversial and in flux.* Julia Sherman, the psychologist who criticized the embedded figures test as a measure of analytic ability in females, also criticizes studies of bilateralization and especially their conclusions about females.[4] She prefers another interpretation of the evidence: Earlier verbal development in females leads them to prefer a verbal approach. They learn by being told and later by reading, which is another form of being told. Thus, simply by preference, girls neglect the nonverbal approach to problem solving. They grow up not only to be more verbal than spatial but, Sherman speculates, altogether to prefer left-hemisphere thinking to right.[5]

The relationship of spatial skills to mathematics is far from clear-cut, however. Some researchers see spatial visualization as needed at all levels of math learning. Others find it high only among mathematicians who specialize in geometry or topology. Many people believe that the most important factor in predicting math

*Other researchers find boys to lateralize early for spatial skills (by age 6) and girls to remain bilateral until later. Where this evidence is cited, the argument is reversed: early lateralization is *better* not worse for spatial visualization. See Sandra F. Wittelson, "Sex and the Single Hemisphere," *Science,* Vol. 193, July 30, 1976, p.425.

aptitude is verbal facility, especially ability in reasoning and logic. Not all high-verbals do well in math, but apparently no one does well in math who is not also high in verbal skills. Even more confounding is the fact that there is not just one kind of spatial ability. Low-level spatial skills (as in Figs. IV-1 and IV-2) may not require any transformation of visual images. High-level spatial visualization (as in Figs. IV-3 and IV-4) is defined as ability to visualize three-dimensional configurations and to manipulate them mentally.

Even though not everyone agrees about the relationship between spatial visualization and math, it is important to deal with this issue. If it can be proved that something about female brain development makes it unlikely that females will enjoy and do well in math, there will be little interest in supporting ventures to cure their math avoidance. Since we know so little about how the brain develops, it seems to me we must deal with the issue in a different way.

Let us grant that the ability to do mental manipulations of visual images is probably useful in learning some parts of mathematics. We have observed that many females do not get the same exposure to spatial visualization training in their play and their home environment. Thus we must try to teach spatial skills to those who need them. Part of the effort being made at Mills College, a women's college in Oakland, California, to attract undergraduate women into higher math is based on these assumptions. Precalculus courses pay particular attention to familiarizing young women with graphing and functions. The students learn to recognize the curves most frequently used in calculus and statistics problems before they study them in advanced courses.[6] Other math educators and specialists in spatial skills are looking for ways to teach these skills to children.

Meanwhile, the field continues to develop quickly. In 1975, in the midst of all this controversy, one study was conducted that might change many of our present beliefs about women and spatial visualization. The study found a group of women who consistently outperformed average men on spatial skills tests. As we have noted, men almost always do better than women on such tests, but in this case, the women excelled. Who were they? Female athletes![7]

There are several possible reasons why athletes do better than average women and men. Perhaps moving their own bodies through space helps athletes improve their sense of distance, and positioning their hands to catch or to return a ball may develop the ability to make quick projections of the ball's flight. We do not yet know enough to draw conclusions from this single piece of research, which, since it has not been replicated, cannot be taken too seriously. But it suggests that spatial visualization can be learned.

Improving Spatial Visualization

Hemisphere specialization does not account for all intuition about spatial relations, and even hemisphere specialization may not just happen. If this is the case, how and where is spatial visualization learned? There is some evidence that participation in shop and mechanical drawing classes increases a student's ability to visualize objects in space. This alone might account for some male superiority over females. Until July 1, 1977, when Title IX of the 1972 Higher Education amendments went into effect around the country, boys were required to take courses in shop and mechanical drawing in junior high school or high school, while girls were not allowed to take these courses and were required to take cooking and sewing instead. This forced specialization is now illegal, but there are still fewer girls voluntarily taking shop than boys. Taking shop decidedly improves spatial ability. Testing before and after the first year of engineering shows that the study of engineering can give the same results. Engineering students improve their spatial skills by studying engineering.

Another source of training in spatial visualization is boys' clubs. Boy Scouts traditionally learn to make and follow maps, to use a compass, to survey, and even to appreciate architecture. They are given courses in "orienteering," or learning to locate themselves in the woods or on a mountain with minimal mechanical aids. Few similar options exist for girls. Girl Scouts may learn knot-tying (in some troops but by no means in all), but not orienteering. So we still wonder whether some boys are more skilled in this intellectual function because they are boys or because they are better trained.

Mapping Your Mind

Folding bed
Diaper
Butterfly net
Bird feather

List as many uses as you can of the objects named above. Use your imagination. The uses do not have to be conventional ones. There is no time limit. Hand in your paper when you are done.

* * *

This is a typical test of "divergent thinking." Students who enjoy an assignment like this one and who find endless possibilities in every object named are considered "divergent" thinkers. Divergent thinkers like open-ended tasks, tolerate ambiguity easily, like freedom to spend as much time as they want on a problem, and enjoy going off in all directions at once.

"Convergent thinking," on the other hand, is directed toward finding an answer or a unique solution. It is usually a process having an obvious beginning, a middle, and an end. "I've found it" means we can stop looking. "I've got it" means we can stop working.

It has been suggested in recent years that part of the reason for dislike of mathematics is that mathematics is perceived to require convergent thinking, the opposite of divergent. Those who do better at collecting information and working it down to a single solution are supposed to prefer math and science to history, English, and even law. But the converse may not be true. In a very important recent study, Lorelei Brush included the convergent/divergent dimension in her survey of attitudes toward mathematics. She found that really good science and math students do both convergent and divergent thinking. Again the issue seems to be ideological rather than actual: people who prefer divergent tasks associate math with a style they *think* is antithetical to their own.[8]

Still, it would be interesting to find ways to teach mathematics to people who like shifting categories of thought, who find many answers to one question, and who produce ideas easily. One starting point might be with the order in which mathematical ideas have been conventionally taught. Teaching math as the ancients and moderns developed it or as a discovery course, allowing students to find out rules and propositions for themselves, as Brush suggests, are two alternatives.*

According to one theory, we have to go through a four-stage cycle in order to learn anything at all.[9] First, we observe something through a concrete experience; then we reflect on what we have observed; we assimilate these observations into a theory, either one that we have thought about before or a new one; and finally we take some kind of action. Ideally, we should have all the abilities needed to observe, reflect, theorize, and act, and the most efficient learners among us probably do. But given our various temperaments and the different learning experiences we have had as children and young adults, we may sharpen our abilities in one or two of these learning areas and underdevelop the others. For example, careful observation with attention to detail is in some senses the very opposite of active decision making and also differs from the ability to sort out information and create a theory. Reflection tends to inhibit action and vice versa. In short, it is very likely that we excel at and enjoy one kind of intellectual activity more than another.

One way to begin to get a picture of our individual learning style is to draw our own "cognitive map." The

*See Chapters Six and Seven for demonstrations of such an approach.

Learning Styles Inventory, available for a nominal fee from a Boston firm, is one way, but by no means the only one, to do this kind of self-analysis.[10]

This inventory asks us to rank ourselves according to some key words like "discriminating," "tentative," "involved," "practical," "watching," "risk-taking," and "future or present oriented." Our scores are then analyzed in terms of the four ways of knowing: concrete experience, reflective observation, analytic-conceptual, and active experimental. Another way to put the scores together helps us see whether we are "convergers" or "divergers."

The goal of cognitive mapping is to gain an awareness of how we think and of differences in cognitive styles. Teachers of different subjects prefer certain learning styles, and very often our preference for a subject may be no more than a delight in the cognitive style we associate with it. Dislike of mathematics, from this perspective, may stem from dislike of the nondiscursive, didactic style of teaching. Conversely, disdain for history and social studies as taught in high school might be the result of having to learn facts devoid of theory.

All such superficial mappings of our minds must be treated cautiously, but it seems better to know something about how we learn before we try to alter our learning style or change the mathematics curriculum to fit our individual needs. Not all learners of mathematics want or need number theory. But some of us might do better if the unit on fractions were preceded by a discussion of the existence of different kinds of numbers. I might not have been so bothered by X^{-2} had I been told clearly and in advance that a negative exponent would not behave like a positive exponent.

For another learner, the same information is best conveyed through discovery. Given some concrete materials to play with, you might learn about dividing fractions entirely on your own and feel more secure about it because you had something to relate it to. Some teachers do try to vary teaching techniques according to the various cognitive styles of their students. But if we are familiar with our own cognitive map, we can control our learning independent of the teacher's temptation to teach a concept as she would prefer to have learned it herself.

Adult learners might decide to start relearning mathematics with books on the history of math or the science of number, as I did. Others may choose to spend several weeks on puzzles to get their heads back into the rhythm of doing math. Still others will do quite satisfactorily with a review book that divides the subject into teachable bits, drills the material, and then tests for mastery.

Since our cognitive maps are forever being reshaped by what and how we learn, none of these preferences will remain fixed. It would be too much to ask any teaching institution to keep a running check of these changes. It is far wiser for us to monitor our thinking ourselves. But to do this we need to be self-confident about learning. We need to be able to say, "This approach does not suit me. Do you have another?" Or "Can I read a book about the calculus before I begin this class?" Or, "Give me an example of how this statistical procedure is applied in a real-life situation. Then perhaps I can figure it out for myself."

The creators of the Learning Style Inventory warn us at the end of the test that the inventory does not measure learning style with 100 percent accuracy.[11]

Rather, it indicates how we see ourselves as learners. Even if at one stage of our lives we cannot learn from concrete experience, for example, it may be because of a childhood experience which, with patience, we can overcome. Mastering something new and difficult can affect our cognitive map in two ways: it can alter the way we think and it can alter the way we think about how we think. If we see ourselves as impatient with detail, unable to recall numbers or amounts, incompetent at one sort of thinking or another, we may well perform or not perform just as we expect simply because we have a "cognitive self-image" as well as a cognitive style to contend with.

This does not mean that all our disabilities are self-imposed. Some experiences in learning mathematics are difficult for everyone, whatever his or her learning style or cognitive map.

Alan Natapoff is a physicist who teaches in the Harvard-MIT division of Health Sciences and Technology and studies the teaching of math at the University of Massachusetts. He theorizes that the human brain does not do very well at a process called groping. He defines exploratory groping as seeking a particular object, solution, or approach in a very large field of possibilities. It is easy to find a pencil, he says, if we are told it is in one of two pockets. We will look systematically in one and then in the other. But if we are looking for lost keys that could be in any one of six pockets, or in a pocketbook, or in the car door, or on any one of four tables or near the shelves, we grope unsystematically: into one pocket and out the other, looking in the same place many times over, getting more and more distraught as we continue our random search. According to Natapoff, math for the beginner, especially word-problem solving, requires just the kind of groping we do worst. No wonder inexperienced math and science students find the process upsetting and unrewarding.

Natapoff's system, which will be described in some detail in Chapter Eight, minimizes fruitless exploration. Lost keys are "placed," for the sake of having somewhere to begin, in one of two pockets. Not that they really are there, but at least the randomness of the search can be controlled by assuming for the moment that they are. Then, if they prove not to be there, the problem is reformulated: "Consider the car door and the door of the house." If this, too, turns out to be unfruitful, "Consider the end tables," and so on. The problem is not solved quickly this way. On the contrary, any experienced key finder would have located the keys long before, using a tried repertoire of key-finding techniques. But at least the beginner can start somewhere and keep track of where he's been. Doing something minimizes the feelings of helplessness and fear.

In fact, Natapoff's system applies quite well to problem solving at the higher levels. Encountering an unfamiliar problem, experienced mathematicians also must grope. Their "Let the basic unit equal passenger-late-minutes-per-mile" is one way to minimize groping. It is a form of "Let us assume that the lost keys are in one of two pockets." Then at least one can go to work.

Is There a Nonmathematical Mind?

> Math, along with related fields like chess and music, has a core of such pure intuition that a child of genius can display his powers quite early, not limited by experience as merely tal-

> ented children are. Instead of soaking up knowledge faster than their peers, prodigies seem almost disconnected from experience.
>
> —T. Branch, "Saul Kripke: New Frontiers in American Philosophy."

Is there a mathematical mind? Yes, of course. At some level of genius connections are made that defy the normal learning pattern and eventually permit the individual to think thoughts no one has ever thought before. One reason for the early display of genius in mathematics is that mathematics can be developed as a system without any reference to reality or to experience. As a young child, Saul Kripke figured out for himself that $(a-b)(a+b) = a^2 - b^2$ just by playing with numbers and noting that sequences like $(7-5)$ multiplied by $(7+5)$ equaled the square of the larger number less the square of the smaller one, $(49-25)$, or 24.

Still, this does not mean that the average person has a nonmathematical mind. For some, mathematical ideas may be self-explanatory; others need repetition and an opportunity to absorb more slowly. For one group of children the numerals 1, 2, 3, and 4 may have more significance than the words one, two, three, and four or the lengths - — —— and ———. From this perspective, mathematics anxiety may occur at the first encounter with another cognitive style or a jump in conceptualization for which the child or adolescent was neurologically or emotionally unprepared. But by the time a student has entered or graduated from college, he or she is not likely to suffer from any of the perceptual difficulties that may once have gotten in the way.

Still, some people, epitomized by the character in the following passage by Philip Roth, are so distracted by what Henry James called "felt life" that they cannot concentrate on the essential mathematical information as it is presented. They are so fascinated with detail, with the "people" part of issues, that abstracting the numbers and the ratios from a problem wrenches them from their real interests. Roth tells of Nathan, a sickly and feverish young boy, whose father tried to sharpen his mind by giving him arithmetic problems to solve. As Roth tells it, the father would announce a problem like this:

> "Marking Down," he would say, not unlike a recitation student announcing the title of a poem. "A clothing dealer, trying to dispose of an overcoat cut in last year's style, marked it down from the original price of thirty dollars to twenty-four. Failing to make a sale, he reduced the price to nineteen dollars and twenty cents. Again he found no takers, so he tried another price reduction and this time sold it . . . All right, Nathan, what was the selling price if the last markdown was consistent with the others?" Or, "Making a Chain." "A lumberjack has six sections of chain, each consisting of four links. If the cost of cutting open a link . . ." and so on.
>
> The next day, while my mother whistled Gershwin and laundered my father's shirts, I would daydream in my bed about the clothing dealer and the lumberjack. To whom had the haberdasher finally sold the overcoat? Did the man who bought it realize it was cut in last year's style? If he wore it to a restaurant, would people laugh? And what did "last year's style" look like anyway? "Again he found no takers," I would say aloud, finding much to feel melancholy about in that idea. I still remember how charged for me was that word "takers." Could it have been the lumberjack with his six sections of chain who, in his rustic innocence, had bought the overcoat cut in last year's style? And why suddenly did he need an overcoat? Invited to a

fancy ball? By whom? . . .

My father . . . was disheartened to find me intrigued by fantasies and irrelevant details of geography and personality and intention, instead of the simple beauty of the arithmetic solution. He did not think that was intelligent of me and he was right.*

The young Kripke was fascinated by the possibilities of number, the young Roth (for this tale must be autobiographical) just as much by the possibilities of personality. To call theirs "mathematical" and "nonmathematical" minds is to miss what they represent: two beacons in a continuum of human curiosity in search of meaning.

References

[1]Ann C. Petersen, "Physical Androgyny and Cognitive Functioning," *Adolescence in Development Psychology, 12,* no. 6 (1976), pp 524–533.
[2]I have not read anything written by Dr. Joseph Bogen but I have heard the slide-lecture he has given around the country. The references are to his lecture.
[3]Robert E. Ornstein, "Right and Left Thinking," *Psychology Today,* May, 1973, pp. 87ff.
———, *The Psychology of Consciousness,* Palo Alto, W. H. Freeman, 1972.
———, "The Rise of Right-Headed People," interview with George Harris, editor of *Psychology Today,* Psychology Today Tape, produced and distributed by Ziff Davis Publishing Co., 1973.
[4]Julia Sherman "Effects of Biological Factors on Sex-Related Differences in Mathematics Achievement," prepared for the National Institute of Education, Contract No. 400-76-0113, Fall, 1976.
[5]Julia Sherman, *Sex-Related Cognitive Differences: An Essay on Theory and Evidence,* (N.Y. Charles C. Thomas, 1978).
[6]The director of the Mills program is Prof. Lenore Blum, Department of Mathematics. She has written a description of her project entitled "Educating College Women in Mathematics: A Report of an Action Program in Progress," based on a talk presented at the conference "Educating Women for Science: A Continuous Spectrum," April 24, 1976. The paper is available

*I am indebted to Helen Vendler for reminding me of this passage.

as part of the Wellesley-Wesleyan Math Packet, for $4.00 from the Wesleyan Math Clinic, Middletown, Conn. 06457.
[7]Marilyn G. Jones, "Perceptual Studies: Perceptual Characteristics and Athletic Performance," in *Readings in Sport Psychology*, H.T.A. Whiting, Ed., Lafayette, Indiana, Balt Publishers, 1972, pp.96ff.
[8]Lorelei Brush, "Mathematics Anxiety in College Students," submitted to the *Journal of Counselling Psychology.*
[9]David Kolb, *A Self-Description of Preferred Learning Modes*, Boston, McBer and Company, 1976.
[10]*Learning Style Inventory: Self-Scoring Test and Interpretation Booklet*, Boston, McBer and Company, 1976.
[11]*Learning Style Inventory: Technical Manual*, Boston, McBer and Company, 1976.
See also David McClelland, "Testing for Competence Rather Than for Intelligence," *The American Psychologist, 28*, No. 1, January, 1973, and Abigail J. Stewart, *Self-Definition: A Personal Style of Cognitive Initiative*, Boston, McBer and Company, 1976.
[12]Philip Roth, *My Life as a Man*, New York, Holt, Rinehart and Winston, 1970, p. 36.

Answers

1. Finish the Box Test. Piece 1 and Piece 5 together finish the box
2. Fold the Box Test. Box 1 is correct.
3. Hidden Figures Test: Figure A is located in Design #4.

5

Word Problem Solving: The Heart of the Matter

de Weese:	How did you know the problem was in the switch?
Phaedrus:	Because it worked intermittently when I jiggled the switch.
de Weese:	Well, couldn't it jiggle the wire?
Phaedrus:	No.
de Weese:	How do you know all that?
Phaedrus:	It's obvious.
de Weese:	Well, then, why didn't I see it?
Phaedrus:	You have to have some familiarity.
de Weese:	Then it's not obvious, is it?

—Robert M. Pirsig, *Zen and the Art of Motorcycle Maintenance*

Given what we already know about the causes of math anxiety, I wish I had the power to change certain things about the teaching of mathematics in the United States. Among them, I would abolish timed tests, by decree if necessary. I would let children use an abacus or calculator to diminish the nervousness that comes from fear of forgetting. I would teach spatial visualization in school, especially to little girls and women; like Pascal's bet on the existence of God, even if it turned out not to matter we would have lost nothing by teaching it. And, not least, I would figure out some way to help people conquer their fear and disability in solving word problems.

Word problems, sometimes called story problems or statement problems, are, in my opinion, at the heart of math anxiety. They appear throughout the elementary curriculum and are the first reasoning problems of any sort that children are given. Since such problems prefigure science problems, students need learning strategies to solve word problems if they are not to grow up avoiding science and math.

More than any other aspect of elementary arithmetic, except perhaps fractions, word problems cause panic among the math anxious. We have seen already how the math anxious refuse even to attempt the Tire Problem and how word problems depress and defeat them. Their attitude, not their math ability, gets in the way.

Lynn Fox asks students enrolled in her math enrichment program what they do when faced with a difficult math problem. Do you stay with it until you have solved it? Do you leave it and return to it later, refreshed? Do you go to someone for help? Or do you forget it? We can classify people by their answers to this question, not because the answers reveal their mathematical aptitude but because they tell us much about their expectations of themselves.

Most difficult problems are not immediately obvious to anyone (even to Phaedrus). There is a good reason why even many capable mathematicians do not like to do math in public. For a period of indeterminate length, one flounders. How well one sticks with a problem through this floundering may well be a function of one's tolerance for floundering in general, or of how well one flounders in math. It would help us to know what mathematicians do when they are floundering. Do they enjoy it? Do they busy themselves with some kind of operation, something between a doodle and a stab at the problem? Do they sketch it? How can one flounder constructively?

At its most destructive, floundering creates a panicky search for some formula that will liberate one magically from the dilemma. Even when confronted by a puzzle without numbers, such as the Cannonball Problem (see below), or by a question that can legitimately be answered in terms of unknowns (like the following), people will seek a formula.

A man earns a monthly salary. At Christmas time, he is given an extra month's salary as a bonus. How much does he earn in that year?

Since no numbers are given, it will be impossible to answer this question in dollars. But the information asked for can be expressed in the terms that are given. There is an unknown: the monthly salary. The employee is given one more of those than there are months in the year. Hence he will earn 13 times his monthly salary, or $13x$; no numbers, no formula, only some degree of recasting of the problem.

Another reasoning problem goes like this:

A tennis tournament has 125 contestants. On losing a match, a contestant drops out of the tournament. The winner must win all the matches he or she plays. How many matches are needed to complete the tournament?

Here again there is no formula. One might be tempted to draw some kind of tree diagram starting with all the players, playing them off one against the other. But if we turn the question around and ask how many people must be eliminated to produce one winner, the answer is easier to find: 124 people will each lose exactly one match. Hence 124 matches will produce a winner.

These are tricky questions in a sense, because we expect to have to do some computation and it turns out we do not.* But, as the psychiatrist Michael Nelson points out, most problem solving mistakes are neither computational errors nor errors in logic. Rather they are psychological in origin. The "I can't" syndrome seems to be particularly disruptive in doing word problems.

How can we learn to handle word problems? One insight, provided long ago by Robert Davis at the University of Illinois, is that when people come up with wrong answers to word problems, they may have the right answer to another question. Therefore, a key to treating fear of word problems is to discover the question the student was answering by analyzing the error he made. We can also do that kind of analysis for ourselves.

The Tire Problem Revisited

Consider the Tire Problem again:

A car goes 20,000 miles on a long trip. To save wear, the five tires are rotated regularly. How many miles will each tire have gone by the end of the trip?**

*Tricky questions are fair; trick questions are not.

**Treatment of the Tire Problem is derived partially from systems developed by Susan Anslander and reported in her "Teaching Word Problems in the College Math Class" (1977, unpublished).

Many people find the Tire Problem difficult, but not because of the mathematical reasoning or the calculations. The notion of rotating the tires is foreign to them. One woman comments; "First I had to imagine what was being talked about, which made me anxious because I wasn't sure I was right. Perhaps there were some rules about rotating tires that I didn't know." Another writes about how a group of adult women dealt with the problem: "Very few people used any common sense with it at all. Some supposed it meant that the car had five wheels. Others thought you had to figure out the rotation or envision some kind of continuously changing set of tires." This is not as unusual as it sounds. Many people get bogged down looking for a mental picture to associate with the problem.

No amount of contempt or condescension can eliminate difficulties such as these. But probing wrong answers can sometimes bring these false notions to the surface, which documents how irrelevant to the solution these issues really are.

Two of the most common wrong answers to the Tire Problem are 4,000 miles and 15,000 miles. The first answer, 4,000 miles, was arrived at by dividing the number of miles traveled by the number of tires, 20,000 divided by 5. This answer is not right, but it can be described as the answer to another interesting and important (if preliminary) question: How many miles will each tire spend in the trunk during the trip? Once that question is correctly formulated, it is only one more step to the right question and from it to the right answer—20,000 miles, the total mileage, less 4,000 miles, the number of miles any one tire will have been in the trunk, is 16,000 miles, the number of miles any one tire will have been on the road.

The second answer, 15,000 miles, comes from a similar realization that each tire will be in the trunk some portion of the time, in this case (incorrectly) ¼ of the time. Thus, it was assumed incorrectly that the tire will be on the car ¾ of the time and ¾ times 20,000 equals 15,000. This answer is not correct because each tire is off ⅕ of the time, not ¼. But the approach was intelligent.

Now consider several ways to get the correct answer, 16,000 miles. One way is to notice that four of the five tires are being used at any given time and to take ⅘ of 20,000 and get 16,000 miles. But this is not the only way. We can think of time in trunk as ⅕ of the time, leaving ⅘ of the time for time on wheels. A third is to reason: If the car goes 20,000 miles, then the tires together drive 80,000 tire-miles. If each of the five tires is used equally, as we have been told, then each tire will go 80,000 tire-miles divided by five tires or 16,000 miles. Note how useful the unit tire-miles turns out to be.

There are sound pedagogical reasons for being flexible with one's students or with oneself. One person's tidiness is another person's mess. One person likes to go from knowns to unknowns; another from unknowns to knowns. One person will start by sorting out the detail, another by getting the whole picture, and others by achieving different combinations of the two. No matter. The key is finding a way that keeps one at the problem because the people who lose are not the ones who are wrong but the ones who give up.

Painting the Room

Problem:

If I can paint a room in 4 hours and my friend can paint the same room in 2 hours, how long will it take us to paint the room together?

The first wrong answer that many people come up with is 3 hours. This is plainly the result of averaging the two amounts of time and dividing by 2: 4 hours plus 2 hours equals 6 hours and 6 divided by 2 equals 3. Logically, this answer does not make sense, though it is computationally correct. If it took both of us 3 hours to paint the room, my friend, who could paint the room alone in 2 hours, would not need my help. Even though the answer is absurd, however, it does show us that the correct answer is going to be less than 2, and this puts us somewhat ahead.*

One constructive approach to the problem is to draw a picture of the room (Fig. V-1).**

On this model we see how much is painted in some

*A math anxious reader responds:

> The fact that an answer is absurd doesn't help me. There are lots of math problems that have people doing things that don't make sense. My sense of reasonableness in regard to math was suspended years ago so an absurd situation won't prevent me from getting a wrong answer.

This reader is reacting to the fact that many word problems are defective. A good word problem should require thinking and logic not merely be a sum wrapped around in words.

**Treatment of the painting-the-Room Problem is also derived from Susan Auslander's "Teaching Word Problems in the College Math Class." Note that it is assumed, though not stated, that painting the room means painting the four walls and not the ceiling too.

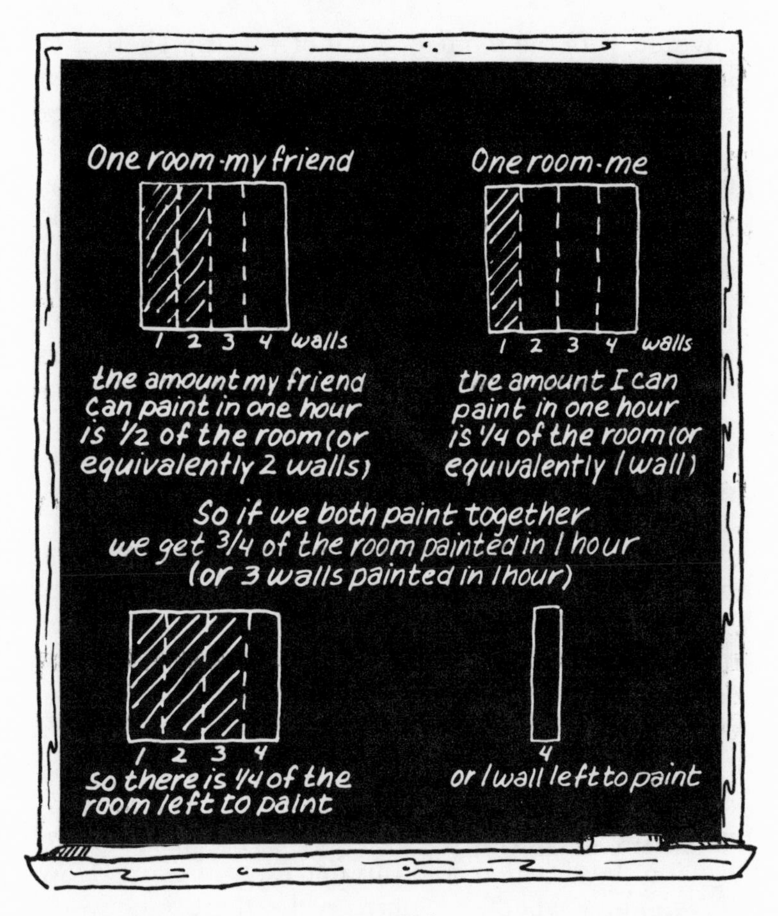

FIG. V-1

This approach to the Painting the Room Problem courtesy of Susan Auslander.

common unit of time by both of us. We choose as a common unit 1 hour. My friend can paint ½ the room in 1 hour; I can only paint ¼ of the room in that time. Thus, at the end of 1 hour painting together, we have covered ¾ of the room (or 3 out of 4 walls). There is ¼ left to do. If ¾ of the room gets done in 1 hour, then the remaining ¼ will get done in ⅓ of an hour. Hence the whole room gets painted in 1⅓ hours. The appeal of this approach is that there is no algebra, no ready-made formula, only a step-by-step working out of the situation. Note also the options: we can consider the room as a whole and think about ½ the room, ¾ of the room, or ¼ of the room. Or, we can think about it wall by wall (see Fig. V-1 again). We can compute the hours and parts of hours, or we can turn the hours into minutes and work with that. We can work out the entire time or choose some unit, 1 hour, and work from there: many ways, many pictures, many solutions.

A less visual way to solve this problem is to say,

Let t = time in hours we spend painting the room together.

This is a way to give the final answer a designation, t, before we know what its value is. The process we follow is algebraic.*

The trouble with an algebra approach for people who are not comfortable with it is that once we have designated the answer, in this case t, we feel as if we are leaving the situation, and this makes us insecure. How do we know whether we are on the right track when we are dealing with fractions and "ts"? One way to feel more secure about the problem, I think, is to stay closer to the information given and not to abstract it quite so soon.

We know that one room can be painted in 2 hours by

*Let t = the time it takes us to paint the room together. In t hours, I paint $t/4$ of the room and my friend paints $t/2$ of the room. $t/4 + t/2$ must equal one room.

one person and that that same room can be painted in 4 hours by another. Adding the hours will get us nowhere because the total time is going to be less than both doing the room separately. Perhaps adding the *rates* at which the two people work is the direction in which we ought to go. But how do we add rates? This is a good question and, as we will see, it takes us to much more sophisticated mathematics.

Let us begin by calling the rates fractions because fractions designate ratios as well as parts of a whole. (See Chapter Six for more on the different meanings of fractions.) My rate is $1/4$ (1 room in 4 hours). It will always be $1/4$ whether I paint 2 rooms in 8 hours ($2/8$), 3 rooms in 12 hours ($3/12$), or 50 rooms in 200 hours ($50/200$). My friend's rate is $1/2$ as it, too, will always be whether she paints 3 rooms in 6 hours ($3/6$) or 50 rooms in 100 hours ($50/100$).

Now, adding the rates, we get $1/2 + 1/4 = 3/4$. So far so good, except that we have forgotten by now what $3/4$ means—that together we can paint 3 rooms in 4 hours. That is our combined rate. But since the question was not how many rooms can we paint together in 4 hours, but how long it will take us to paint 1 room together, we have to change that rate from 3 rooms in 4 hours to 1 room in so many hours. The way we do that is the nub of the operation.

Well, we can say, "Three rooms in 4 hours changed to 1 room in so many hours," several times until we get an idea of how to proceed. (Not a bad way.) Or we can express this, too, as fractions:

$$3/4 = 1/?$$

If at this point we notice that the numerator (3) has been divided by 3 to get the new numerator (1), then remembering the rules about fractions, we might consider dividing the denominator (4) by 3 to get the new denominator, $\frac{4}{3}$. This is the right answer ($1\frac{1}{3}$ hours), but it is not the only way to get the answer. Another way, once we have found the combined rate ($\frac{3}{4}$), is to divide it into one room:

$$\frac{1}{\frac{3}{4}} = \frac{4}{3}$$

and arrive at $1\frac{1}{3}$ hours, too.

This problem seems difficult to do because it *is* difficult. Adding or comparing rates is not a simple matter. The calculations may be simple, but the principle behind them is quite advanced. So if the Painting-the-Room Problem gives us trouble, it is not because we are dumb but because we are smart enough to intuit that there are complex ideas just beneath the surface.

Currents and Speeds

Another rate problem produces a similar difficulty.

A ship goes in one direction, west to Hawaii, at 20 nautical miles per hour and, because of the wind, makes the return trip at 30 nautical miles per hour. What is its average speed?

The first time they hear this problem, many people will react to it as to the Painting-the-Room Problem and average 20 nautical miles and 30 nautical miles to arrive at an "average speed" of 25 nautical miles per hour. The mistake here is to assume that both trips took the same number of hours. But in fact, both trips covered the same distance, but not in the same time. Take any distance just to check this out. For a common distance of 600 nautical miles, going one way (600 ÷ 20) would have taken 30 hours; going the other way (600 ÷ 30) would have taken 20 hours. Thus, the round trip would take 50 hours and 50 hours divided into 1200 nautical miles gives an average speed of 24 knots*. Again we are trying to compare rates, and the way we compute average rates, as we would have learned if we had gotten to statistics, is different from the way we compute averages.**

Although the Ship Problem is not a calculus problem, comparisons of rates of change are the kinds of problems found in mechanics and dynamics, out of which the calculus developed. In all these problems the calculations are simple once we have formulated the problem correctly, but formulating the problem is not at all easy.

Another example is the problem of upstream travel against downstream currents.

If a boat travels in still water at 6 miles per hour and the current is 3 miles per hour, how long will it take the boat to go 10 miles?

The computation is deceptively simple. Actually we just subtract to get the rate per hour (6−3=3) and divide the total miles (10) by the rate per hour (3).† But

*A neat formulation for this is:

$$\frac{D}{20} + \frac{D}{30} = 2D \qquad 30D + 20D = 600$$
$$50D = 1200$$
$$D = 24$$

**Averages are computed, as most of us know, by adding up the elements and dividing by the number of items added. Average rates are computed by taking the reciprocals of the numbers, $\frac{1}{20}$ and $\frac{1}{30}$ in this case, adding these, and dividing them *into* the number of items, in this case:

$$\frac{2}{\frac{1}{20} + \frac{1}{30}} = \frac{2}{\frac{3}{60} + \frac{2}{60}} = \frac{2}{\frac{5}{60}} = \frac{2}{\frac{1}{12}} = 2 \times \frac{12}{1} = 24$$

†See Answers for Chapter Five.

the person who likes to think through and understand such a problem may have a harder time, because it is hard to figure out how fast or slow the boat is going at any one instant. Does it go 6 miles an hour and then get set back? Or does it somehow go backward and forward at the same time? It is not easy to form a mental picture of what is going on. Movement is very difficult to visualize. Indeed, it can only be indicated with arrows if one tries to depict it at all.

Non-numerical problems are sometimes even more difficult to handle than the ones that can be reduced to numbers. Take the problem:

If we were to lay dominoes on a checkerboard so that each domino covered two adjacent squares, how many dominoes would we need?

Many people will simply go blank when confronted by such a problem. For one thing, they don't have a clue to how many squares, black or white, there are on a checkerboard. They cannot even begin to think about the problem. If we are told that there are 32 black squares and 32 white squares and we know that each domino is to cover two adjacent squares, then we can figure out that there will be 32 dominoes on the board. Mathematics helps here not by providing any formulas or systems of calculation, but by suggesting a way to organize knowns and unknowns. But it only helps if we are on the right wavelength to begin with.

Cultivating Intuition

Obviously problem solving is not really a matter of making logical deductions from memorized formulas, but an exercise in imagination. Cognition has sometimes been defined as "seeing relationships." It is that and more: it is fantasizing about relationships, trying out imagined ones until one is found that fits the situation. The use of tire-miles in the Tire Problem is a good example of imagination at work. The relationship that links all the tires and all the miles driven has to be construed out of nothing and provides an interesting way to get control of the problem.

No one can consciously control the image-making part of the brain, the faculty of intuition or insight. Beginners and people who have not done math problems for a long time come to believe that because they do not have instant flashes of insight they have no intuition. Since it is always easier for the teacher, the tutor, or the text to provide one image for the learner to apply than to wait for the learner to develop her own, some people never even find out that they can invent images for themselves.

Mathematical intuition is often mistaken for that mystical "mathematical mind" so many people are persuaded they do not have. Actually intuition can be developed like any other skill. It responds to exposure to math and to other related experiences. Phaedrus' "intuition" about the faulty switch, quoted at the beginning of this chapter, was nurtured in long hours of fixing electrical connections. What was obvious to him was not at all obvious to someone who had not had the same experience.

An example of seeming linguistic "intuition" illustrates this point. The beginning German student with little background in other foreign languages and a not very sophisticated knowledge of English will have to memorize the distinction between the words *abreisen,* to travel away from, and *anreisen,* to travel to. Other students, just as new to German but already familiar with Latin, may intuit the difference without much effort, but that is because they already know that in Latin *ab* usually means "away" or "from" and *an* means "to" (abnormal, announce, annex etc.).

The inexperienced beginner, to take another example from German, may have trouble translating the phrase *Im Laufe des Tages* ("in the course of the day") even when he knows the meanings of the words *Tag* (day) and *laufen* (run). A student who is acquainted with French or Latin, knows a little Italian from home, or has a more sophisticated knowledge of English, however, will probably guess correctly that in this phrase *Laufe* does not mean "running" but "course".* Just the same, what seems like a better language facility is only more experience with language.

Mathematical intuition at the levels we are discussing may not be so different from other kinds of intuition. The more experience one has in solving problems and the more varied one's mathematical repertoire, the more facile and intuitive one will appear. To cultivate our mathematical intuition, then, we must collect and keep fresh lots of pieces of information and many kinds of strategies. Then, when we need to, we can quickly search among a rich store of ideas for those that will help solve the problem at hand.

*As in the French *courir,* to run, or *au courant* for being on top of things, etc.; or the English "current" for running water.

The Cannonball Problem

What kind of experiences improve and exercise mathematical intuition? Consider the Cannonball Problem.

Imagine that you have twelve cannonballs, all but one of equal weight. You are not told whether the odd cannonball is heavier or lighter than the others, only that it is different. You are given a balance scale and told to try to find the odd cannonball. To do this problem well, you must find the odd cannonball in three weighings.

This problem is good for precisely the reason that kept me from solving it. There is no one trick. Rather you are forced to move from a fairly general approach to a more and more systematic one as you get more deeply involved.

Let's try it.

With or without pencil and paper we note fairly quickly that weighing six cannonballs against the remaining six does not reveal anything much. (See Fig. V-2.) One side goes up on the scale; one goes down. But which side has the odd ball? Since we do not know whether the odd ball is heavier or lighter, it could be in either group.

Some people may try weighing five against five, but it soon becomes obvious that it will take more than two weighings to proceed from five unknown balls to one. Weighing four against four turns out to be more fruitful though, as we will see, not nearly as fruitful as we first assume.

Figure V-3 shows the results of weighing four against

Fig. V-2

four and the possibilities that emerge from this first weighing. Either the two sets of balls balance, as in Fig. V-3A, in which case we know that the odd ball is in the remaining set of four; or one side goes up and the other down, as in Fig. V-3B, in which case we can dismiss the set of four not weighed and concentrate on the two sets we have weighed.

In mathematical terms, we are treating the twelve cannonballs as three sets of four balls each. The first weighing tells us that the odd ball either lies in the set we have not weighed or in one of the two sets we have weighed. At this point we have either an odd ball in a set of four (in mathematical terms, four unknowns) or an odd ball in one of two sets of four (in mathematical terms, eight unknowns).

At this stage, the tendency is to forget about the "normal" set or sets and concentrate on the one or two sets that contain the odd ball. But this turns out to be a poor strategy, because the "normal" set or sets have known balls that can be weighed against unknowns.

FIG. V-3 A

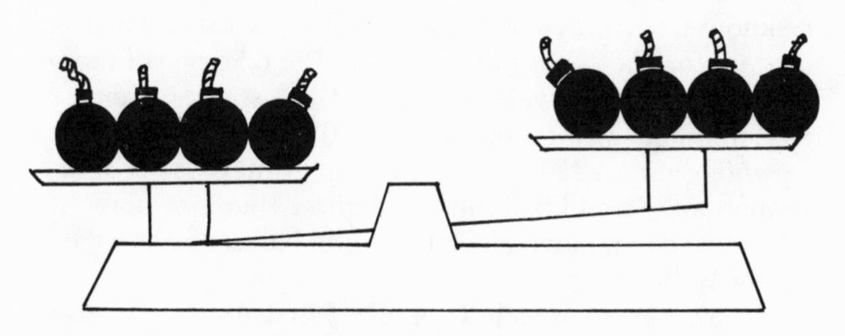

FIG. V-3 B

More of this later.

Let us pursue some of the implications of the first weighing. We might think that finding the first two sets of four of equal weight would very much simplify things; after just one weighing we have located the odd ball in a single set, as one of only four unknowns. The problem is that not knowing whether the odd ball is heavier or lighter than the normal balls, we will not get far enough fast enough even with this lucky contingency. Weigh two against two and the scale will go up on one side or the other. That doesn't tell us whether there is a light ball on one side of the scale or a heavy ball on the other. Thus, the only way to cope with four unknowns—*as far as we know at this point**—is to measure one against one, twice, and even then we won't know which one is odd.

The way we got to this dead end is instructive. Since the unknowns alone give us information only very slowly (if at all), it might be useful to save the knowns from the first weighing and to use them to measure the unknowns. A set of unknowns weighed against a set of knowns, for example, will indicate at the very least whether the unknown set is heavier or lighter than normal. Figure V-4 illustrates this second weighing. We know from previous weighings that two sets balance; therefore both have normal balls. We take one of those normal sets and weigh it against the last set and either the last set will go up on the scale as in Fig. V-4A, in which case we have a new and important piece of information: the odd ball must be lighter than normal or it will go down in which case the odd ball is heavier. (See Fig V-4B.)

This is about as far as I got on my first attempt to solve the problem. I had had one insight: not to ignore the sets of balls that turn out to be normal. I had used that insight to figure out whether the odd ball was heavier or lighter. At this point I thought I had broken the problem, but I could not find the odd ball among four

*There is a way to solve with four unknowns in two weighings, but it requires a step we have not thought of yet.

FIG. V-4 A

FIG. V-4 B

unknowns in only one weighing—and I had only one weighing left—even though I had the heavier-or-lighter issue resolved. So at least another insight was called for, if not another approach.

At this point, people who are not bored or frustrated

may change perspectives and look ahead to where they want to be by the third weighing. What will I need to know and what is the maximum number of unknowns I can have left if I am to solve the problem in the third (last) weighing? Four unknowns are too many, at least using the technique of comparing sets. But three unknowns are not too many to sort out in one weighing if the heavier-or-lighter issue is resolved. Figure V-5 shows why.

If we weigh two of three unknowns in the third weighing, only two results are possible: either, as in Fig. V-5A, one ball goes up and the other down, which tells us which one is odd (since we have already found out that the odd ball is heavier or lighter); or, as in Fig. V-5B, the two balls balance, which tells us that the third ball is odd. So the second insight, based on an examination of the end of the problem, is: by the third weighing, I may have no more than three unknowns and I must know whether the odd ball is heavier or lighter.

Where are we now? We have a fruitful first weighing (four against four) and a set of conditions for the third weighing. So we have to focus on the second weighing. This is the tough one. Indeed, long after I had worked on this problem, a math instructor at a nearby high school recalled for me that he had spent eight days and eight nights working on the Cannonball Problem when he was in high school eighteen years before; and of that intense week of work, six days, he said, were spent trying to figure out what to do with the second weighing!

At this point I had to concede, with great reluctance, that the division of the twelve cannonballs into sets of four had been inadequate and that I had to find some way to identify the cannonballs individually. Having to

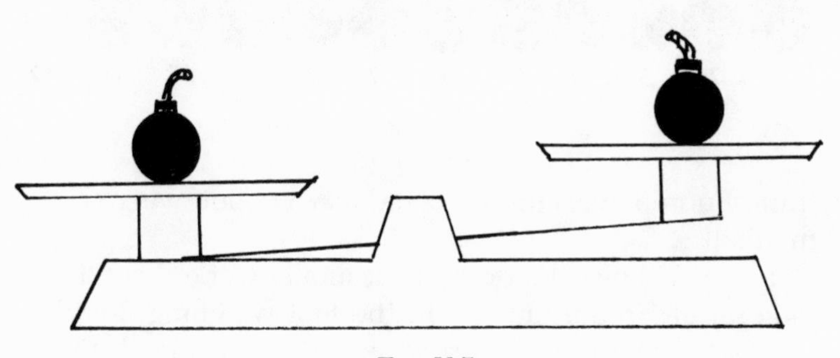

FIG. V-5 A

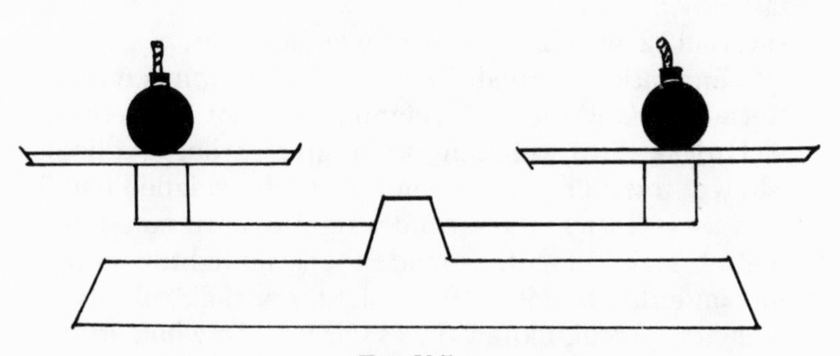

FIG. V-5 B

start the problem over again is just the sort of experience I like least. But there was no other way out. So I began again, this time with cannonballs numbered 1 through 12 (Fig. V-6).

I began to weigh again, four against four, etc., but this

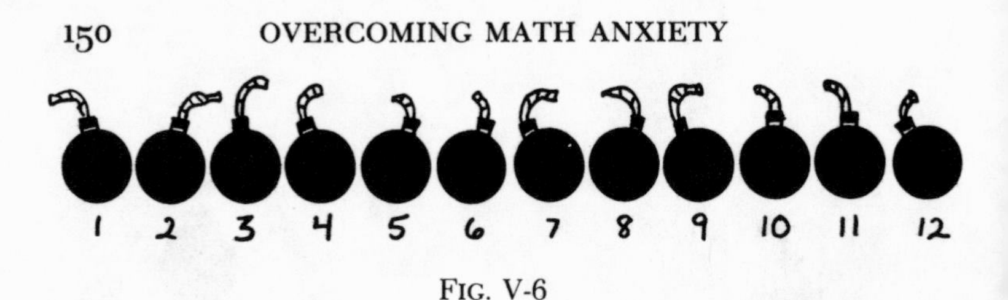

FIG. V-6

time I noted everything I could learn about every cannonball.

Fig. V-7 shows the problem as finally worked out (for one set of circumstances). In the first weighing, I took two sets of four. As it turned out, they balanced (Fig. V-7A). So in weighing two I selected three balls (9, 10, and 11) from the remaining four and one ball (8) from the known group of normals. Weighing these in sets of two, I came up with the result shown in Fig. V-7B: balls 9 and 10 together raised the scale, balls 8 and 11 lowered it (Fig. V-7B). Since the scale was out of balance already, I knew my odd ball was one of three: 9, 10, or 11. Ball 12, which I had never weighed, was, by process of elimination, normal. Of course ball 8, from a normal set weighed in the first weighing, was not in question.

On the third weighing, then, given what had been shown in the first and second, I simply weighed ball 9 against ball 10. In the second weighing I had noted that balls 9 and 10 together made the scale lighter; so this meant either that 9 or 10 was lighter or that ball 11 was heavier. (Again, I knew 8 was normal.) So when, in the third weighing, the scale on the side of ball 9 was lower and the scale on the side of ball 10 was higher, I had my odd ball and I knew its weight: ball 10 had lightened the scale twice, in weighing two and in weighing three (Fig. V-7C). No other ball had done that twice.

To solve the problem satisfactorily, of course, one

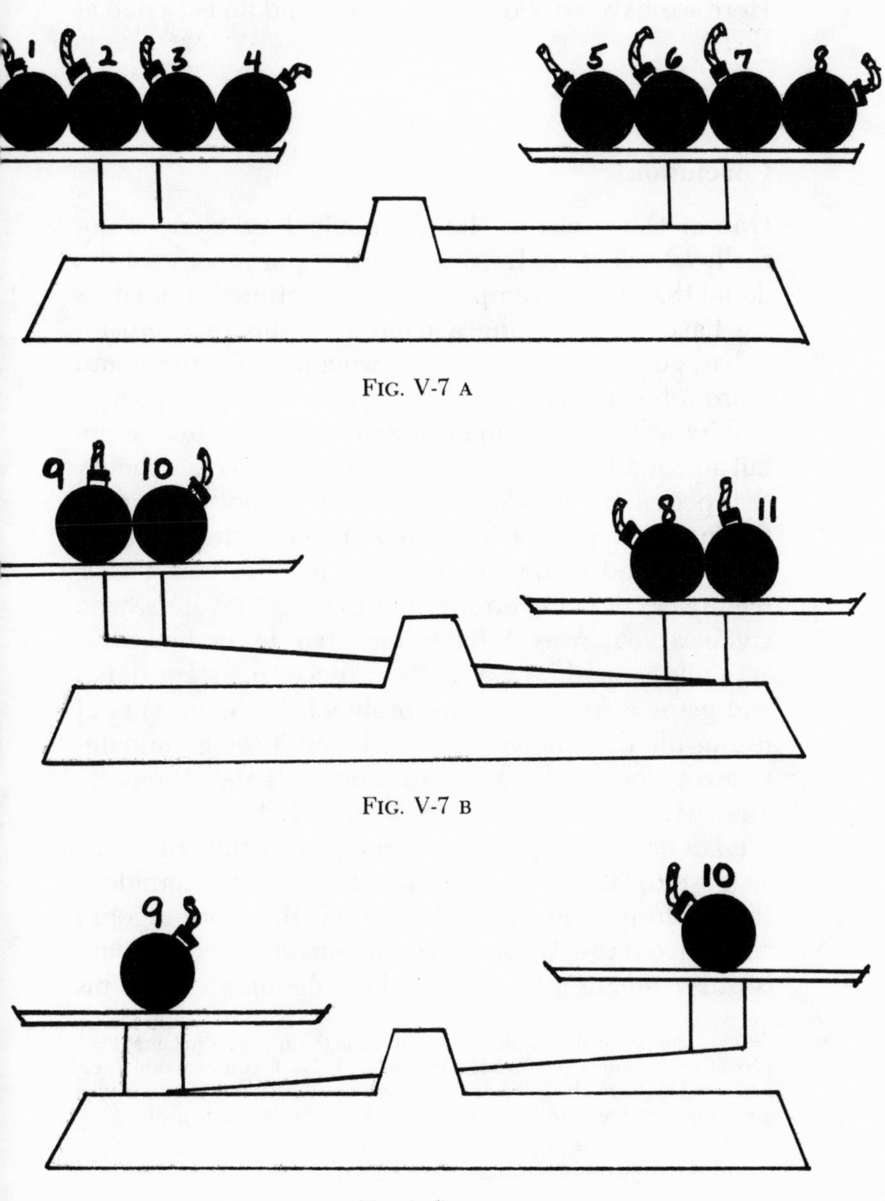

FIG. V-7 A

FIG. V-7 B

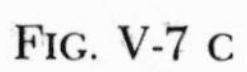

FIG. V-7 C

must do it in three weighings for all contingencies. Here we have considered only one, and an easy one at that.*

Conclusion

One of the tricks in the Cannonball Problem is not really a trick at all. It involves attending to some of the detail that we are tempted to ignore at first. Indeed, as we have seen, getting a handle on this problem involves going back to previous weighings for more and more information.

Why will one person at one moment pass over a detail and another focus on it? Psychiatrists have a theory that people do not like detail if they cannot manage it. Or, in a more popular version of the same theory, there are supposed to be two kinds of people: sharpeners, who make detail even more visible, and levelers, who try to carve it away. What is important for us, however, is to understand how a mathematician manages detail and gains control over the problem. Is the process of listing all the knowns and unknowns, of giving unknowns letter designations and relating them to knowns a way of managing detail? "Let t = the time it takes us both to paint the room," one problem solver suggested; "then . . ." and she went on more confident than before. One theory has it that the word algebra comes from the Arabic word *algebrista* which in Spanish also means bone-setter. The algebraist, with his

*The solution for all contingencies appears in the appendix to James Fixx' *Games for the Super-Intelligent* (Doubleday, 1977). Fixx uses another version of the problem: the goal is to find the one counterfeit coin among eleven good ones; but the solution is of course the same as for cannonballs.

transformations and manipulations, can "set" a problem into a form that allows him to solve it.

From this perspective, then, panic in the face of apparent ambiguity might be just another way to feel out of control. Yet detail, instead of muddling up the issue, could provide reassurance, if one could attend to it constructively. Some people to whom I have given the Cannonball Problem numbered the balls long before they had to. They felt better dealing with controlled detail. I, a leveler, hoped I could avoid such systematizing. I figured that since I was looking for one ball and would be eliminating several with every weighing, I could get away with not attending very closely to the others. Sometimes, very often in fact, my system saves me a lot of time, but sometimes I get caught!

Perhaps the stickiest part of the Cannonball Problem is the incomplete information we have about the odd ball.* If we knew from the outset whether it was heavier or lighter, the problem would be easier to solve. Yet it is not obvious at the beginning of the problem that this information is going to be so critical. On the contrary, at the beginning we think that the difficulty is going to be in sorting out the eleven from the one. When the other issue becomes critical, we either

*To know when one has enough information to solve a problem is a problem in itself. Take this one, which a group of mathematicians took just about as long to solve as anyone:

> Two women meet after many years. One asks, "How old are your three daughters?" Answer: "The product of their ages is 36." Question: "But that's not enough information." Answer: "Well, the sum of their ages is the same number as the post office box that we shared at college." Question: "But that's still not enough information." Answer: "The oldest one looks like me." Question: "Oh, now I know their ages."

If you can figure out why that last remark gives the questioner enough information, then you can figure out their ages. See answers for Chapter Five.

retreat from the problem in anger or disgust, or dig in even more deeply and recommit ourselves to working it out. What one does at this point is as much a matter of temperament and experience in solving problems as it is a matter of ability or something as global as gender.

Partial information is by no means peculiar to mathematics, but, as we have seen, some people have a lower tolerance for "ambiguity" in math or science than in poetry, history, or foreign languages. Is this because they have already learned to distrust math and to doubt their ability to extrapolate from partial information to the knowledge they need? Or can they be taught to cope with partial information in every field?

Some educators are beginning to teach problem solving (heuristics) independent of any particular field or discipline. Others think that certain exercises like sketching or working with concrete materials can effectively ready people for problem solving in math. It is by no means certain that math problem solving can be taught in the same way as facts or interpretation of material. But there is no doubt in my mind that facility in solving word problems can be increased with practice. In the math clinic I have been associated with, people are advised to do word problems every day, like morning exercises, to add to their repertoire, to increase their exposure to all types of problems, and above all to reinforce their self-confidence.

A woman I know set her family to do the Cannonball Problem one evening. One (male) child turned away from it almost immediately; another son worked hard for an hour, then threw it over in disgust. Her husband, a scientist, enjoyed it and solved it that evening. Her second son might have solved the problem had his focus not been disrupted by failure.

One of the most obvious differences among individuals in attacking the Cannonball Problem is their willingness or unwillingness to keep at it, to pick it up after an interval and try again even if an apparent solution turns out to be false. Some people's attention is disrupted by failure. Others have just the opposite reaction: the joy of mathematics is precisely the challenge of it, the feeling that it is there to come back to, and the knowledge that there is no need to finish it in one sitting. Failure, or lack of immediate success (which is quite another way to think about failure), bothers them not at all.

The question is of course whether, owing either to innate and ingrained characteristics or role socialization some people more than others are less likely to develop the appropriate temperamental characteritsitics for solving word problems; or whether the experience of success itself makes people more tolerant of failure.

Whatever the cause of inadequate problem-solving ability, the issue is whether people who do not have it can develop the "right temprament" for doing math.

Maybe there is no "right temperament" at all: Mathematics has some standard moves, many rules and facts and proofs. But as I observe myself and others doing math, I notice that we can make ourselves aware of our own perceptual and strategic preferences, the kinds of models we prefer, the way we approach problems best. Thinking about problems leads inevitably to thinking about thinking. In the Tire Problem, one person likes to get a picture of the problem in terms of the smallest unit, the number of miles one tire has gone during one cycle, or the number of miles one tire will be in the trunk. Another will invent a unit to help him get the whole picture. Take the painting-the-room problem. One person will translate it from hours into minutes to avoid dealing with fractions. Another prefers hours and even fractions of hours because big numbers make her nervous.

Such individuality could be acknowledged and encouraged, although it plays havoc with lesson planning. In mountaineering, climbers learn to move in terms of both the mountain and themselves. Strong arm pull, good or poor natural balance, preference for chimneys, and tolerance of exposure on the mountain are considered in deciding where and how to move up. Learning to do math, like learning to climb, involves above all learning about oneself.

Answers

1. The man's annual salary that year is 13 times his monthly salary, or $13x$.
2. The boat is traveling at 3⅓ miles per hour.
3. A good way to solve this problem is to lay out the possibilities after the second question has been answered in a format like the following. What groups of numbers multiply to 36 and how do they add up?

Possibilities	*Product*	*Total*
$6 \times 6 \times 1$	36	13
$9 \times 4 \times 1$	36	14
$18 \times 2 \times 1$	36	21
$3 \times 2 \times 6$	36	11
$3 \times 3 \times 4$	36	10
$12 \times 3 \times 1$	36	16
$9 \times 2 \times 2$	36	13

We must assume that both women remembered the post office box number. If it had been 14 or 21 or 11 or 10 or 16, the problem would have been solved when the answer to the second question was given. Since it was not enough information, the post office box number must have been 13, which meets both the multiplication and the addition requirements and "is not enough information."

Two possibilities remain that both add up to 13: 6,6, and 1 and 9, 2, and 2. When the first woman says, "The oldest one looks like me" she is giving it away: the oldest one is not a twin. Hence the combination is not 6, 6, and 1 but 9, 2, and 2.

6

Everyday Math

Socrates' best known treatise on government does not begin at the beginning. Rather it starts with an idea, the notion of justice, which is somewhere in the middle of the subject. Many who have written about the Socratic method of teaching have identified Socrates' questions as the method's quintessential characteristic. It is true that he asked his colleagues to start with what they knew and got them to think in response to his probing. But I do not believe that the uniqueness of his method lies in the dialectic alone. It is rather that he abandoned the conventional way of thinking about a subject. He started somewhere, anywhere at all, much as he might throw a stone out into a river and then systematically pursued the ripples it produced. Thus the discussion of justice, for example, moved into the subject of appropriate tasks and appropriate rewards. We might not agree with the definition of justice as it finally emerges from the *Dialogue*, but we have learned a lot about government from the progress of the debate.

In the same way, the sequence in which elementary mathematics is learned may not be the only way to study it. Instead of starting at the beginning, wherever that might be, we could start somewhere in the middle, with an interesting question that inevitably will bring us around to the beginning, or at least to other interesting mathematical ideas. Adults, whose conceptual equipment is already fairly sophisticated, might best learn elementary mathematics the second time around by diving in somewhere, anywhere at all, and, assisted by an informed interlocutor, proceeding in ever-widening concentric circles.

To demonstrate this approach (which is, incidentally, the one I use), I have selected four topics, three in elementary mathematics and one fairly advanced, to treat not as steps in a hierarchy of accumulated knowledge but as points of interest along the way. I assume, as Socrates did, that my readers are experienced, able to think for themselves, and anxious to understand. In this chapter the way is somewhat familiar because most of us have studied these topics before. The material in Chapter Seven will be new to many.

The topics I have selected interest me. They are: fractions, the many meanings of minus, averages and averaging, and (in Chapter Seven) the calculus. One probably cannot go very far in mathematics without learning computation skills, but perhaps one needn't begin with them. Because of this belief, I am prepared to upturn the normal sequence in learning mathematics.

Let's give it a try.

Fractions the Second Time Around

> Arithmetic is usually taught as all scales and no music.
>
> —Persis Herold

Why is it so difficult for adults to remember or figure out how to go from a fraction to a percentage? This topic comes up again and again in workshops for math anxious people. "How do you get a percentage out of $7/16$?" "How do you know what to divide into what when you are told that a man pays so much in taxes, at such and such a rate, and you need to deduce his total income from that information?"

I have been wondering about these questions for months and, in the process, teaching myself fractions all over again. My conclusion is that fractions are not at all easy; it might have been wiser to postpone learning fractions until we needed them in algebra. I have also seen that specialists in math education and mathematicians themselves strongly disagree about how to present fractions, even how they should be defined. I have frequently written in one draft what one specialist has told me about fractions only to have another reader

© 1977 United Features Syndicate, Inc.

mark up the sentence with "No, not at all true." Again, as always, we will do best to get our clues from the consumers, people who have had difficulty with fractions.

Division of fractions is the problem, as the math anxious adult remembers it. The rule "to divide fractions, invert the divisor and then multiply" simply went over people's heads. Most of them sailed through addition, subtraction, and even multiplication of fractions without ever having to think about what fractions *meant.* When they got to division of fractions, the fact that they didn't know what they were doing finally got in the way. Most math educators agree that we do not have to understand fractions thoroughly to manipulate them, but we have to have a thorough understanding of fractions to feel secure about using them.

What would this thorough understanding involve? Well, for one thing, a simple fraction can indicate any one of four (count 'em) complementary ideas:[1]

First, ¼ can mean one-quarter of one, that is, a *certain part of a whole* (Fig VI-1).

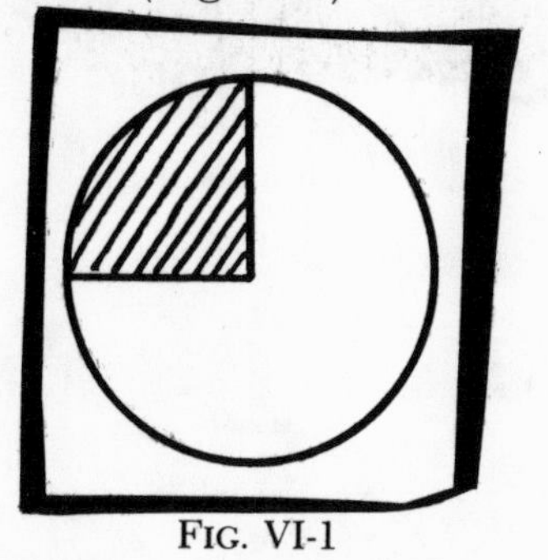

FIG. VI-1

But ¼ can also refer to one item of several, that is, a *certain part of a group,* as in Fig. VI-2.

FIG. VI-2

Third, ¼ can refer to a comparison or a ratio: one item *compared* to four others, as in a one-for-four batting performance in a game (Fig. VI-3).

And finally, ¼ represents the result of a division; in this case the fraction means "four divided into one" (Fig. VI-4).

In the rate problems in Chapter Five we needed a way to express painting the room in so many hours and we (correctly) used fractions: one room in four hours was expressed by ¼; one room in two hours was expressed by ½; and the combined rate, the fraction ¾,

FIG. VI-3

FIG. VI-4

told us that three rooms could be painted by two people working together in four hours.

As I remember, the only picture I had in my mind as I was learning fractions the first time was Fig. VI-1, the pie picture. It would have helped me to be given the idea that fractions meant ratios, proportions, and statements of division as well.

Let us examine Fig. VI-4 a little more closely. If ¼ means "1 divided by 4," then going from a fraction to a decimal or a percentage is nothing more (or less) than extending the meaning of the fraction—not a *change* in the fraction at all. I think this is one of the great truths people who have trouble with math never get. The process of "transforming" 7/16 into a percentage is not a transformation at all, but just another way to state the divisional nature of the fraction.

If we divide 4 into 1 (¼) or 16 into 7 (7/16), which is what the fraction *means* to begin with, then we cannot fail to do the percentage transformation correctly. There is no need to ask, "What gets divided by what?" We know it must be the denominator (the lower figure) divided into the numerator (the upper figure) no matter what the size or complexity of the fraction. In the case of ¼ (1 divided by 4) the decimal equivalent is approximately .25 (approximately 25 percent). In the case of 7/16 (7 divided by 16) the decimal equivalent is approximately .44 (approximately 44 percent). This should work just as well when the fraction is made up of other fractions, such as the fraction

$$\frac{\frac{3}{4}}{\frac{2}{3}}$$

I prefer this notation myself to the more common

$$3/4 \div 2/3$$

because it reinforces the divisional nature of the entire operation—things are being divided by elements which in turn are being divided by other elements.

I have asked many people whether and if so when they began to understand all these varied usages of fractions and to realize that when considering a problem the first step is to decide which aspect of the fraction is going to be useful. Most of my contemporaries cannot remember ever being told this but I notice that people who succeeded in math managed to figure it out at some point and that students who had trouble with math did not. Perhaps this long misunderstood idea—that there are different meanings for fractions—makes otherwise competent people feel insecure with fractions and percentages.

It doesn't help much that fractions are generally introduced to us in terms of pies. So long as we have one pie to be divided among six people, we can see that there will be six portions. This model works well for multiplication of fractions (indeed it illustrates very well why multiplying fractions may result in smaller amounts), but it does not help us at all in visualizing division of fractions. In division we want to know not how many *portions* there are but how many *times* one portion of the pie will go into the whole. Many of us have gotten stuck right here, not on the computation but on the *shift in the meaning* of the answer from *portion-of-pie* to *times.** See Fig.

*Division of whole numbers is a more complex concept than most of us realize. There are two ways of looking at division. Six divided by three means "How many threes are there in six?" But it also means "If we were to divide

VI-5 for a more useful way to illustrate division of fractions.

Let us go back to the rule "invert the divisor and then multiply" to see whether we can make more sense of it now. Our problem is: ¾ divided by ⅔, or as I like to write it out

$$\frac{\frac{3}{4}}{\frac{2}{3}}$$

Our goal will be to eliminate the fraction under the line because when we have done this we will have

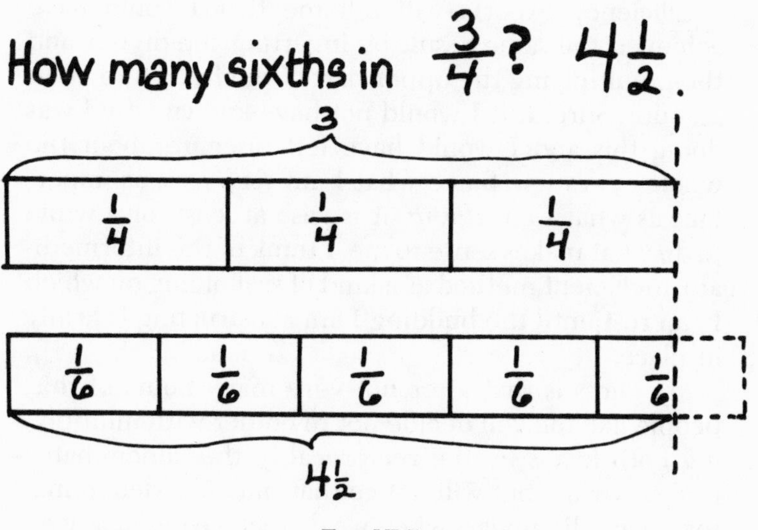

FIG. VI-5

---

six into three equal parts, what would be the size (or count) of each part?" Think about these meanings for a while until you can call each one into view at will.

completed the division.

How will we eliminate the lower fraction? One way is to multiply it by its reciprocal—a fancy word for the fraction that is its opposite. The opposite of $2/3$ is $3/2$. Obviously, when we multiply $2/3$ by $3/2$ we get $6/6$ or 1 and this will nicely take care of the lower fraction. But remembering the rule that what you do to a fraction below the line you must do to a fraction above the line, we will also have to multiply $3/4$ by $3/2$.

$$\frac{\frac{3}{4}}{\frac{2}{3}} = \frac{\frac{3}{4} \times \frac{3}{2}}{\frac{2}{3} \times \frac{3}{2}} = \frac{\frac{9}{8}}{\frac{6}{6}} = \frac{\frac{9}{8}}{1} \text{ or } \frac{9}{8}$$

Efficiency experts will tell me that I could have achieved the same result by inverting the divisor and then multiplying the upper part of the fraction by that amount. Sure. But I would not have known why I was doing this and I would have felt insecure about the whole operation. Since what I am *feeling* is as important as what I am *doing,* it is wise at least for a while to do what makes sense to me. I think of this intermediate, inefficient method as a kind of scaffolding on which I can rest until the building I am constructing is firmly in place.

My guess is that after not very many hours of this, people like me will decide not to bother with multiplying both levels by the reciprocal of the denominator (the new rule) but will notice that since the denominator always disappears when we do that we might as well shorten the process into two steps:

1. Find the reciprocal of the denominator.
2. Multiply the numerator by that reciprocal.

But when the decision is made to cast off the scaffolding, it will be *our* decision. We will be ready for the short cut and, most important, we will know why we are taking it. In the process, we will also have learned something important about fractions and their reciprocals; that when a fraction is multiplied by its reciprocal, the product is 1. And that insight will be useful to us later on.

Now let's look at what causes trouble in mastering decimals or "decimal fractions," as they are sometimes called. Persis Herold, who has written an extraordinarily useful book for teachers and parents anxious to help their children do elementary arithmetic, suggests that confusion with decimals may derive in large part from their layout.[2] Fig. VI-6 shows what the various places around the decimal point are supposed to mean.

One space to the right of the decimal point is the "tenths" area; but the "tens" are *two* spaces to the left. *Two* spaces to the right of the decimal point is the "hundredths" realm, but the "hundreds" place on the other side is *three* spaces to the left. Herold has noticed that young people who have difficulty with decimals assume—and didn't we all?—that the decimal point is the midpoint of the number, a sort of fulcrum around which the tens and tenths and hundreds and hundredths are deployed.

But where is the midpoint? Look at Fig. VI-6 again. It is the area of the ones! No wonder we make mistakes in reading decimals and feel uncomfortable with them.

When we want to multiply or divide a decimal by ten or some multiple of ten, we are told to "move the decimal point." This is not at all what we are doing: rather

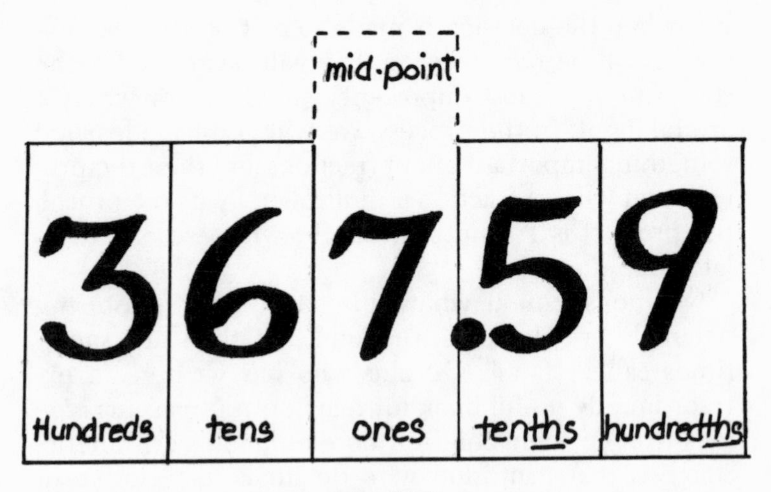

FIG. VI-6

we are moving the ones and tens and tenths and hundredths to different positions around the decimal point. The decimal 165.7 multiplied by ten equals 1,657. It may appear that we have "moved the decimal point" (to the right), but it is more accurate to say that we have moved the 7 from the right-hand side of the decimal point to the left-hand side of it. The decimal point, in theory and in practice, too, should stay put.

Though decimals can be converted into fractions, not all fractions are decimals. Fractions can have any denominator: $^1/_7$, $^7/_{16}$, $^{16}/_{64}$, and $^{16}/_x$ are all fractions. Decimals have only powers of ten as denominators, and percentages, which are a subgroup of decimals, have only one denominator, 100. Thinking of fractions, decimals, and percentages this way prevents this kind of error:

Problem: 6 is what percent of 24?
Wrong answer: 6 is 4 percent of 24.

Four percent is $4/100$, not $1/4$. The right answer is $1/4$ or 25 percent.

Reviewing fractions this way raises the question: What makes us think something is hard or easy? One reason we may think something is difficult is that we learn it *later* in school; that is, we take sequence very seriously indeed. Since we learn to subtract before we learn to multiply, we assume multiplication is harder. Yet subtraction is harder to do than multiplication. We may think some one thing is harder than something else because of what teachers, parents, and peers *say* about it. The calculus was presented to me as "real mathematics" with such emphasis that I was frightened even to try it. Sometimes a concept is straightforward but its notation is very complex. This holds true in statistics, and even some algebraic notation may cast such a long shadow that an otherwise fairly clear idea becomes murky. A fourth reason, the only one that should matter, is that the concept may actually be difficult to comprehend.

When we relearn elementary mathematics, it is well to sort out for ourselves what is apparently or "ideologically" or notationally difficult from what is conceptually difficult. Take the following kinds of percentage problems, for example:

1. What is 25 percent of 24?
2. 6 is what percent of 24?
3. 6 is 25 percent of what?

It is estimated that 80 percent of all people questioned can solve the first problem; about 40 percent can solve the second; and perhaps only 10 percent can solve the third, especially if the numbers are not as simply related as the ones I have chosen here.* One reason why the third problem gives more trouble than the first two may be psychological: the problem seems more open-ended. In the first two problems, the largest number is known and this puts a limit on the range of estimation.

My conclusion from reviewing fractions is that fractions are difficult because they are yet another number system added on to what we already know. Rarely are all these number groups presented clearly, as in Fig VI-7, and in relation to one another.**

Fractions behave quite differently from whole numbers and produce different results when used in familiar operations. Integers multiplied by integers produce larger integers. Fractions multiplied by fractions produce (generally) smaller fractions. These changes do not mean that the meaning of the word "multiply" has changed, as we once thought, but that the numbers are a different kind. In each case we need to know what kind of number we are dealing with and which rules will apply. Furthermore, while we may know without much reflection that eight is bigger than seven, we do not always know simply by looking at fractions that ¾ is bigger than ⅔. (One of the advantages of decimal notation is that the size of the

*Unfortunately I cannot track down the source of this estimation. It has been repeated to me without citation. Although it cannot be taken as fact, people who have read this book in manuscript form say it "seems correct."

**From *The Math Teaching Handbook.* See reference section for citation.

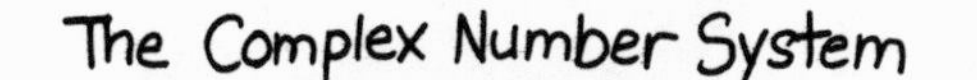
The Complex Number System

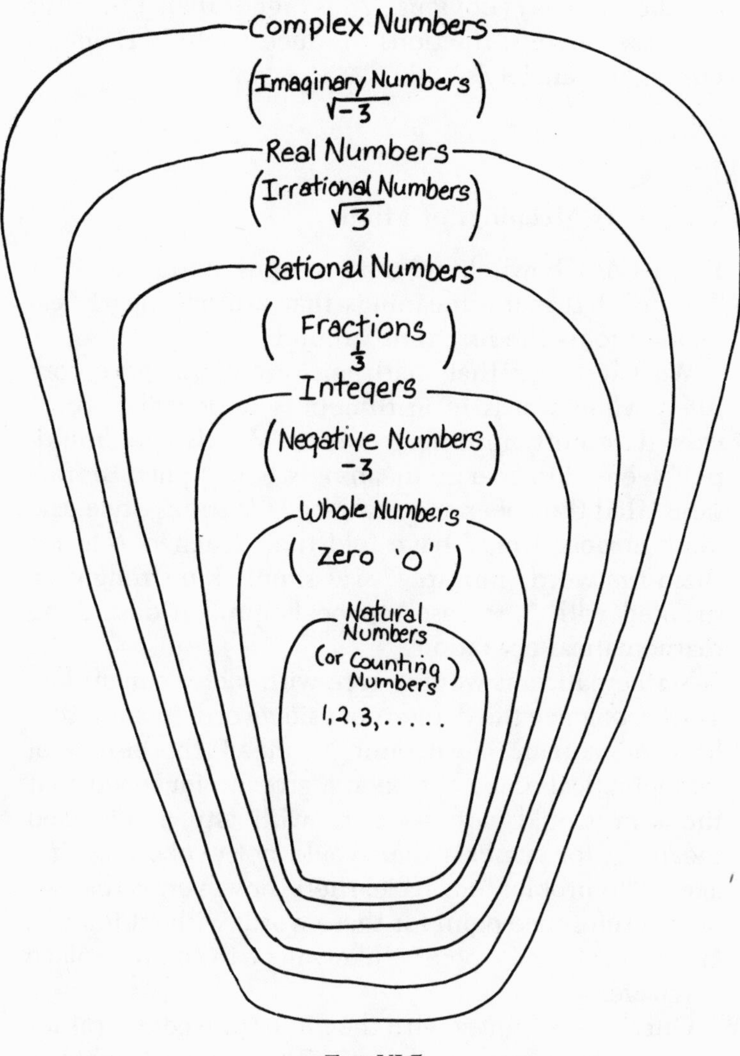
Complex Numbers
(Imaginary Numbers √-3)
Real Numbers
(Irrational Numbers √3)
Rational Numbers
(Fractions 1/3)
Integers
(Negative Numbers -3)
Whole Numbers
(Zero '0')
Natural Numbers
(or Counting Numbers)
1, 2, 3, ...

FIG. VI-7

number is always obvious: .75 is bigger than .667.) For all these reasons, fractions produce confusion, uncertainty, and anxiety.

The Many Meanings of Minus

If fractions bewitch us, minus signs betray us. For "minus" has many meanings that are not clearly explained to us the first time around.

We have seen that math anxious adults were confused when words in arithmetic sounded like words used differently in other contexts. Words like "multiply" seemed to change meaning when applied to fractions. Had they been more aware of learning style, the math anxious would have told their teachers that for them the word "multiply" was simply too strongly associated with "increase" to be helpful in describing mathematical operations.

Mathematicians would agree with such a conclusion. They recognize the dangers of using words in math that have more than one meaning. Indeed, the history of mathematical usage has been a groping for terms that the user would *not* associate with any unintended meaning; for symbols that would not evoke false images. The problem for us learners, however, is that we need a reference point for these words without images; and that reference point is likely to be in commonplace language.

Our first encounter with the minus sign comes rather early in elementary arithmetic. The subject is subtraction and the format is the following:

$$\begin{array}{r} 10 \\ \underline{-4} \\ 6 \end{array}$$

To make this operation come alive for little children, teachers often use potentially misleading phrases like "take away." This encourages the pupil to think of subtraction as removing one quantity from another.

Subtraction is the first and best known meaning of the minus sign, but subtraction is a lot older than the sign itself. The minus sign only came into common usage in the fifteenth and sixteenth centuries as a way to simplify and standardize mathematical notation. Before then, other signs were used. The capital letter M, for example, stood for "minus" in English and German, but of course it might have made no sense to the Chinese.

The problem is that the minus sign (−) does not designate subtraction alone. It has at least two other meanings in elementary mathematics, and although some pedagogues have suggested that we change the signs to reflect these different meanings, the minus sign is still used for all three.*

The second use of the minus sign is to designate a negative number. In my day, negative numbers were actually called "minus numbers," to make it all the more confusing; today the preferred term is "negative number." There are several ways to understand the idea of a negative number. One can *approach* the idea via subtraction: $10 - 4 = 6$; $10 - 14 = -4$. This tells us how to get to -4, but not what it means.

Another approach is to visualize a thermometer (Fig VI-8).

*Mathematicians suggest "−" for subtraction; a high minus sign for the negative number, "$^{-}10$"; and a small degree sign for opposite, "$^{\circ}x$".

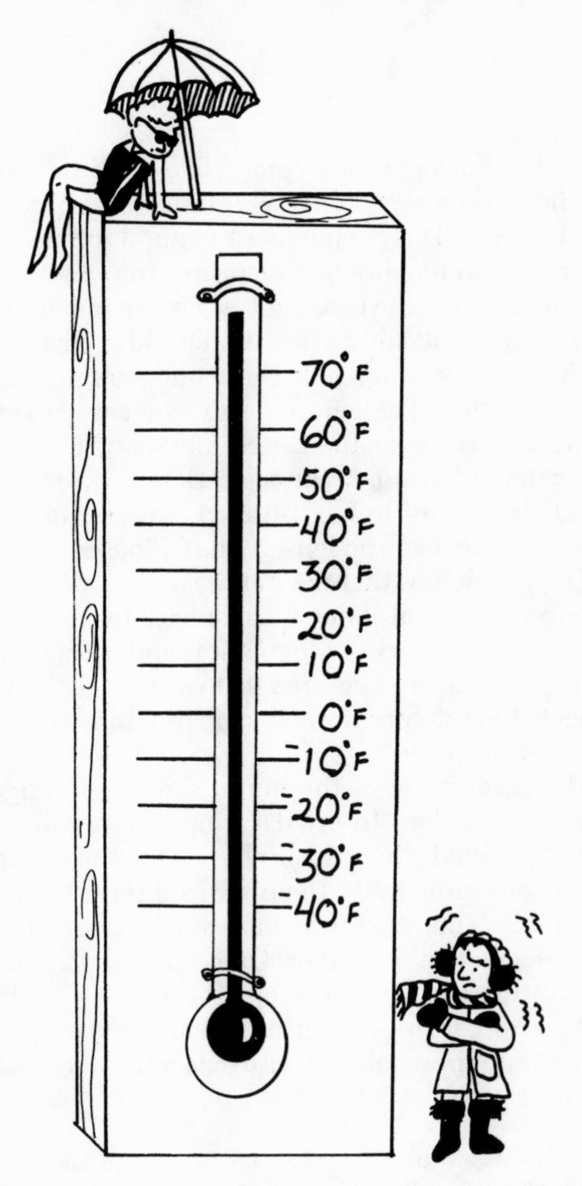

FIG. VI-8

This helps us see that a negative number is on the other side of zero and further that it is a number in a different *direction* from positive numbers. Another model, the East-West diagram, makes this idea of direction even more concrete (Fig. VI-9).

Another way to grasp the meaning of a negative number is to think of overdrafts in banking. If we cash checks for more money than we have in the bank, we can accumulate a negative account. The overdraft model is particularly useful in that it makes sense of the idea of *adding* negative numbers: -10 *plus* -4 equals -14.

From this perspective, adding and subtracting signed numbers might better not be called "adding" or "subtracting" but be conceived rather as a kind of combining operation. We can imagine ourselves moving backward and forward, up and down, or sideways (depending on the visual model) as we combine nega-

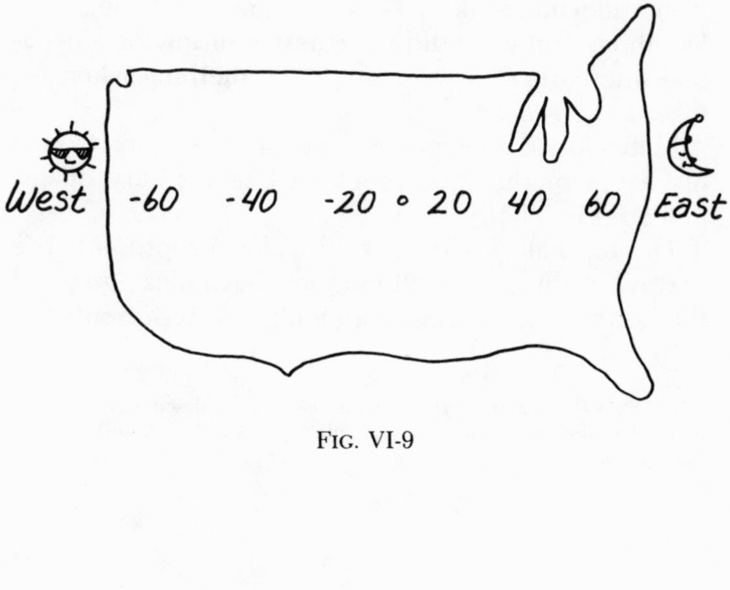

FIG. VI-9

tive and negative, negative and positive, or positive and positive numbers. It would then be easier to understand that addition does not *necessarily* result in an increase in value; the operation is rather an instruction to "go up" or "go East;" similarly, subtraction no longer means "remove" or "take away" but rather "go down" or "go West." The minus sign has gained another meaning.

Things are considerably harder to imagine when we want to multiply negative numbers. It is one thing to add overdrafts and quite another to multiply them. Let's follow this through.

The rule for multiplying (or dividing) signed numbers goes something like this: "Like signs produce positive products; unlike signs produce negative products." By inference, then, there is no operational difference between $(-5) \times (+2)$ and $(+5) \times (-2)$. Yet when I was learning about signed numbers, since I need to visualize, these two operations seemed very different indeed. Multiplication was a shorthand for addition. I could well understand taking (-5) two times or taking (-2) five times. But what did it mean to multiply *by* a negative number? What do we *do* to something when we take it -2 *times?*

I have found an explanation in an illustration from a history of mathematics book that seems very quaint indeed (Fig VI-10).*

The top half (9 o'clock to 3 o'clock) represents the positive realm, where all numbers have plus signs; and the bottom half (3 o'clock to 9 o'clock) represents the

*Lancelot Hogben, in *Mathematics in the Making,* describes this model. Pinhead Studios and I have created this particular illustration from that description.

FIG. VI-10

negative realm, where all numbers have negative signs. We are to imagine all mathematical calculations beginning at 3 o'clock (another convention), but the number of turns around the clock is different for like and unlike signs.

If you cannot deal with that model, try the picture my father uses, reproduced in Fig. VI-11. But even this, helpful as it is reminding us what to do, does not tell us *why* a negative multiplier acts the way it does.

If you are confused, fret not. My only purpose in describing the clock model is to show how very forced the "explanation" is.* This is because there may be no

*This explanation has not been much improved on in four hundred years. In the summer of 1977, a reader of the *Harvard Magazine* wrote to the

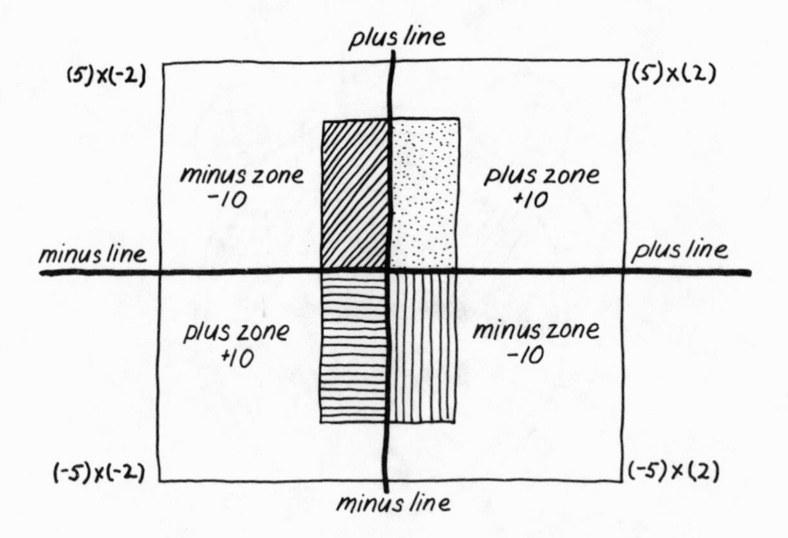

FIG. VI-11

common sense explanation for the behavior of negative numbers in multiplication; the only explanation is that, for arithmetic to remain consistent as a logical system, certain operations must produce certain results.

Once we understand that the system is designed for our convenience, we may not be as disturbed by it as we were before. In arithmetic, the multiplication of positive and negative numbers must produce negative numbers even if the meaning—in our sense—remains obscure. Again, we have to suspend disbelief because

---

column, "Any Questions?" for an expert answer to the question: Why does the multiplication of two minus numbers produce a positive number? The most useful part of a long and complicated answer was the following from John T. Tate, professor of mathematics at Harvard, who said, "There is no natural law that decrees that the product of two negative numbers has to be a positive number. This rule is simply a convention that is universally agreed upon by people working with numbers because it is useful."[3]

we are not in a realm where the measure of truth is reasonableness. In math the test of all reasonableness is consistency!

Still another use of the minus sign is in the expression $-X = 6$. Negative X may look like just a negative number, but it isn't, because X is not a number. When $-X = 6$, $X = -6$; the minus sign here is telling us to "change signs." Thus, in one sense, this minus means "opposite." One person has suggested that for $-X$ and similar expressions, the minus sign be replaced by an "°" and $-X$ written instead as $°X$. This notation would avoid an entirely different use of the minus sign, but (with a few exceptions) it has not taken hold so we are left with a third use of the minus sign, quite different from the other two.

My problem with X^{-2} might have been eased had I known of this meaning of minus. In this case the negative indicates an inverse, that is that X^{-2} has an "opposite," namely X^2, which when multiplied together equals 1. In the same way, X^2 has another "opposite" which when multiplied times X^2 equals 1. That opposite is $\frac{1}{X^2}$. Thus, it is less surprising that the two "opposites" of X^2 will be the same. The following is by no means a "proof." But it is surely a reassurance:

X^{-2} times X^2 = (signs are added when we multiply) X^0 or 1.

$\frac{1}{X^2}$ times $X^2 = \frac{X^2}{X^2} = 1$

A further cause for confusion is that in elementary school, when we encounter these apparent inconsistencies, we are not told that a symbol may have more than one meaning. The equal sign symbol itself has two meanings: that the quantity on one side equals the quantity on the other (as in $6X = 64$) and that the name of the thing on one side is the same as the name of the thing on the other one (as in $X^{-2} = \frac{1}{X^2}$). Had someone shown me this distinction between identity and equal quantity years ago, I might have had less trouble with mathematics. Without knowing that it existed, I needed the three-dash equal sign, "≡," which means "means the same as" or "by definition." Using this sign would have made my life easier.

It would be misleading to leave this discussion without some recognition of the enormous benefits of mathematical notation, especially modern algebra. Fig. VI-12 shows many ways of writing simple equations just during the two-hundred-year period of the Renaissance.[4]

Note the number of languages used as well as the variety of symbols. One man, Vietà, decided to make the vowels stand for the unknowns and the consonants

Development of Algebraic Notation

date	mathematician	symbolism used	modern form
1494	Pacioli	"Trouame .1.n°. che giõto al suo q̄drat° facia .12."	$x + x^2 = 12$
1559	Buteo	1◇P6eP9⊏1◇P3eP24	$x^2+6x+9=x^2+3x+24$
1577	Gosselin	12LM1QP48 aequalia 144M24LP2Q	$12x-x^2+48 = 144-24x+2x^2$

FIG. VI-12

for the knowns. Descartes chose the first letters of the alphabet for the knowns and the last letters for the unknowns, a convention still in use today. Modern mathematicians use all the Greek letters and many other symbols, too. Although it does make learning more difficult, the abbreviated notation of modern algebra makes all mathematical operations easier to do. We can see from Fig. VI-12 how difficult it would be to manipulate equations if the notations were still so cumbersome.

The adoption of generalized symbols was especially important because it freed thinkers to examine and manipulate the operations and relationships they created, whether or not these were realistic or possible. So long as the ancients did not use generalized formulas, they were limited in what they could consider.

Imagine for a moment a time when negative numbers (-4), irrational numbers ($\sqrt{7}$), improper fractions ($^{11}/_{9}$), or imaginary numbers ($\sqrt{-2}$) had not yet been conceived.* Then the following equations could not have been solved:

$X + 6 = 4$ (because X would equal -2)
$2X = 5$ (because X would equal $^{5}/_{2}$)
$X^2 = 7$ (because X would equal the square root of 7, $\sqrt{7}$)
$X^2 = -4$ (because X would equal the square root of -4, $\sqrt{-4}$)

Expressing these same equations in general terms, however, by substituting letters for numbers, allows us at least to consider solutions:

*This example is taken from Tobias Dantzig, *Number: The Language of Science* (see Reference Section for citation).

$$X + b = a \qquad X = a - b$$
$$bX = a \qquad X = \frac{a}{b}$$
$$X^n = a \qquad X = \sqrt[n]{a}$$

The advantage of this generalized notation is that within it one can work even on "illegitimate," "meaningless," or even "nonexistent" solutions. This frees the thinker from irrelevancies, enabling us to consider possible relationships before we have any experience with them. In this way, notation is like language, for we have the words "red" or "father" or "democracy" in our minds before we have any particular experience with them. Generalized notation permits the kind of supposing that led Newton to wonder whether things on earth were not naturally at rest, but naturally in motion, and if they are in motion, why they should ever stop. Astronomers could predict the existence of the planet Pluto before it was ever observed through a telescope and physicists could predict the existence of certain nuclear particles long before they were isolated, because they bore a mathematical relationship to particles already known to exist.

We are left in a quandary. Do we name impossible things with impossible words, thus avoiding any confusion between the language of math and the language of everyday life? Or do we choose familiar language that is loaded with content? The minus sign is an interesting case in point.

Averages and Averaging: One Way to Think about Statistics

Nothing is as useful or as intimidating as statistics. The "fact" that "three out of four people" do or do not do;

that "40 percent of the labor force" is or is not; that "the typical American male" earns or wants or believes, is very familiar rhetoric today. Our president's staff includes his own polltaker to keep him informed of what the public is thinking. Commercial polling has become a large and prosperous industry.

We are intimidated by statistics, not only because they seem to denigrate our intuition but because those of us who are uncomfortable with figures believe that statistics is a science of numbers and relationships in which human judgment is unimportant. In fact, statistics is no more and no less than a substantial improvement on simple averaging. However complex and computerized the techniques may be, reliable and useful statistics must depend heavily on human judgment.

Take averages and averaging as a case in point. We can add up some measurable instances, say the heights of a group of children, the sizes of a group of objects, the incomes of a group of families, or the amount of June rainfall during the past four years. Then dividing that sum by the number of children, things, families, or Junes, we can find the "typical" height, size, income, or June rainfall. Averages give us a sense of things, a feel for the central tendency or the trend.

But averages have severe limits, too. The hardest thing to do in dealing with averages it to get out from under the generalizations they suggest. Once we have a statistic, we must still decide whether the information it contains will help us. This calls for judgment based on our knowledge of the situation and the purpose for which we are gathering the data. It may require us to calculate other statistics before we can decide whether to use the information we have; or it may require us to use another technique to test the reliability of our "facts."

Having added all the heights of children in the second grade at a particular school, can we predict the size of next year's second graders accurately enough to order new chairs? How useful would it be to know, for example, the average income of people living in a metropolitan area? If we add their incomes and divide by the number of families, we will get an arithmetically correct answer, say $12,000. But that might not be quite so useful if we know in advance that there is a large number of rich people who earn considerably more than $12,000 and an even larger number of poor people who earn less. In fact, we may not want to know "average income" at all, but how many people earn more or less than some amount. Depending upon the use of our statistic, it might be more informative to know that half the population earns about $7,000, a quarter earns $10,000, and the richest quarter earns around $25,000.

Different uses call for different techniques. If our goal in this study of incomes is to understand how a particular family compares to all other families in the population, we will want still another kind of average, a picture of how far or near this family is from the midpoint if all families were ranked in order of income.

In addition, we need techniques to deal with unreliable data. We must make sure that our heights, rainfall, number of children born, or reported income are typical or even reasonable. Averaging is based on the assumption that if enough incidents are gathered, some typical pattern will emerge. If we use unusually high or low examples because the sample we chose to measure had some oddballs in it, our picture will be distorted.

Here, too, human judgment is at work. We approach the data with a certain expectation of what is typical. We have already calculated the average, non-numerically, in our minds. If the average comes out high or low, we examine the data to see if atypical examples were counted. We cannot automatically throw these out. After all, they are data. But if we can prove to ourselves that they are there because of an unusual circumstance, and here again our judgment comes into play, then we can dispose of them.

One such example occurred after the mandatory deposit on bottles and cans was made a law in the state of Oregon a few years ago. The state government decided to check thirty segments of highway on a regular basis to see how much, if any, less litter was being thrown out of car windows. The non-numerical expectation was of course that people would save their bottles and cans for deposit redemption. But would highway litter show a decline? The segments were checked, the information put on computer and averaged. The data indicated that twenty-five of the thirty segments showed between 60 and 90 percent reduction in the amount of litter; two showed a reduction of between 20 and 40 percent; one showed an increase of 1 percent; and two had increases of 250 percent.

The last two segments were so out of line that an average of litter per segment might have misled the policy makers. Possibly something had changed in these locations over the year.* But in order to generalize from the sample of segments, the segments have to

*The unusual circumstance in this case was that two overzealous college students, hired to check segments, were counting orange peels and other degradables as litter. The lesson here is that the criteria must be clearly spelled out.

be representative. Statisticians have a way to deal with this kind of problem. Using formulas lay people do not have to employ, they measure the data's confidence limits, which indicate how much confidence one ought to have in one's data.

Sometimes one wants to know not just the average but the amount of variation around the average as well. It was not very meaningful to know that the average income for a family in a metropolitan area was $12,000. We might want to know the range or what is technically called the "interval", too, from $3,000, the lowest income, to $35,000, the highest. We might even want to know something more: how frequently various incomes occur. This kind of work is at the heart of social science: finding the "frequency distribution" and the "amount of variation from the mean (or average)." This is so important because we need to know whether our limited picture is likely to predict the next set of segments, or the larger population, or the next day's weather.

What we really want to know is how tall are the next three second-grade classes likely to be? And will we need to have higher-backed chairs? If the last thirty second graders have averaged four feet but in an interval between three feet six inches and four feet six inches, then the average alone does not give us the whole story. But if we include information about the range and frequency of the deviation from the average, we may be able to order our chairs with confidence.

So far we have been talking about static averages: snapshots of the typical American family or a typical June rainfall pattern or the size of a typical second grader. But our questions may require us to predict trends that change over time, moving averages, things that alter course as other elements affect them. To find the amount of energy required to heat or cool a typical home in Wendover, Utah, for example, the average temperature for the year will be misleading. It is about 51.9 degrees, which would require little heat and no air conditioning. Even knowing that the interval is between 28 and 75 degrees does not help much. The picture of energy requirements has to be a moving snapshot, with averages by the month or even by the week, lined up one after another.

But how lined up? In Fig. VI-13, note that Eureka, California, with an average temperature of 52.2 degrees, ranges only from about 45 degrees to 56 degrees over the year.

Sometimes moving pictures are expressed in bar graphs, sometimes in line graphs, sometimes as curves. The distinction between bar graphs and line graphs is that bar graphs usually represent discontinuous change

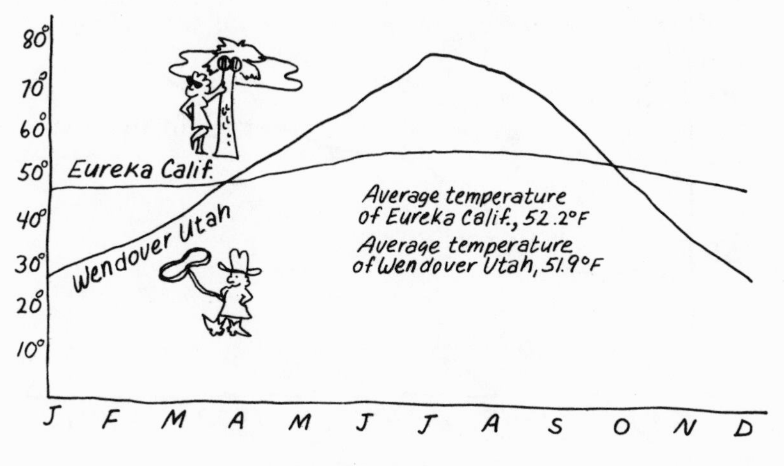

FIG. VI-13

and line graphs usually represent continuous change. The static picture in a continuous sequence is written as a point, with all the information about it contained in the location of that point on a graph. The moving picture is a series of points, connected with a straight or, if need be, a curved line. If no line can be drawn because the points are so scattered, then another technique is used to find out what line most approximately links these points.

Say we wanted to get a picture of how the American people are doing financially, compared to how they have done in the past. To do so we would need to look at a number of different moving pictures.

Figure VI-14, representing the Consumer Price

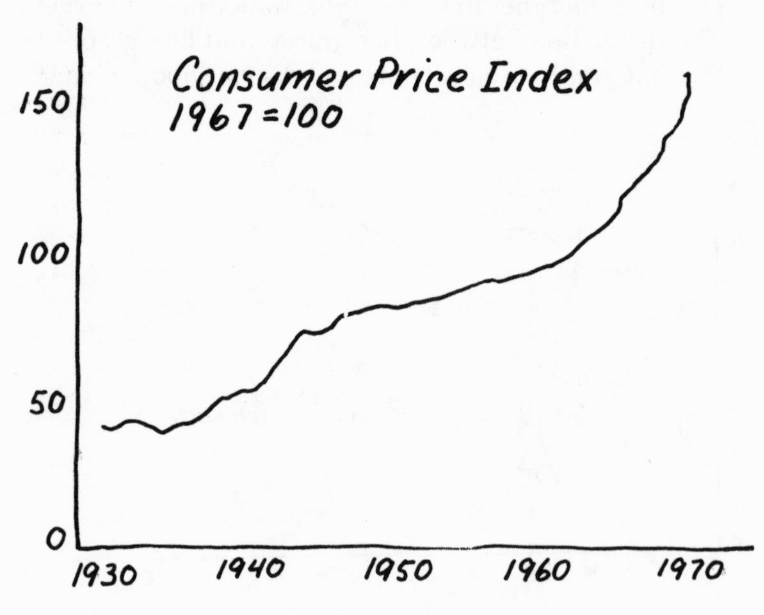

FIG. VI-14

Index from 1935 to 1977, is an interesting place to start. It is quite obvious that the curve of prices goes up rather steadily after the effects of the Depression are past. Note also that just for the sake of having some criterion, the year 1967 is selected as equaling 100; not 100 anythings in particular, just 100, so that we can call a low point 45 and a high point 155. Here we want comparisons, not exact numbers, so the arbitrary designation of 1967's prices as 100 is perfectly all right.

The Consumer Price Index would, by itself, suggest that things are not going well at all, for the price of goods has increased 300 percent since 1935. If hourly wages had stayed the same, then it would take three times as much work to buy goods in 1977 as it did in 1935. We also need to know how incomes behaved in this period to get a feel for the average pocketbook today. Figure VI-15 is helpful in this connection. Note that the wholesale price is given not in absolute or even in relative numbers, but in terms of the purchasing power of $1. In the last graph of prices, the curve went up. In this one, the curve goes down. Why? Because we are measuring different aspects of the same phenomenon: in the one the price itself, in the other the purchasing power of the dollar. These two numbers behave like reciprocals; smaller numbers have larger reciprocals. (Seven is larger than 2; $\frac{1}{7}$ is smaller than $\frac{1}{2}$.) Figure VI-15 tells us that the purchasing power of the dollar in 1967 terms went down slowly until 1970, when it began a very steep decline that continues to the present. This graph gives us a rather vivid picture of the effect of inflation on our purchasing power.

Still, we cannot be sure how we are doing unless we know how incomes fared in the same period. Figure VI-16 gives two approximations. The solid line gives a

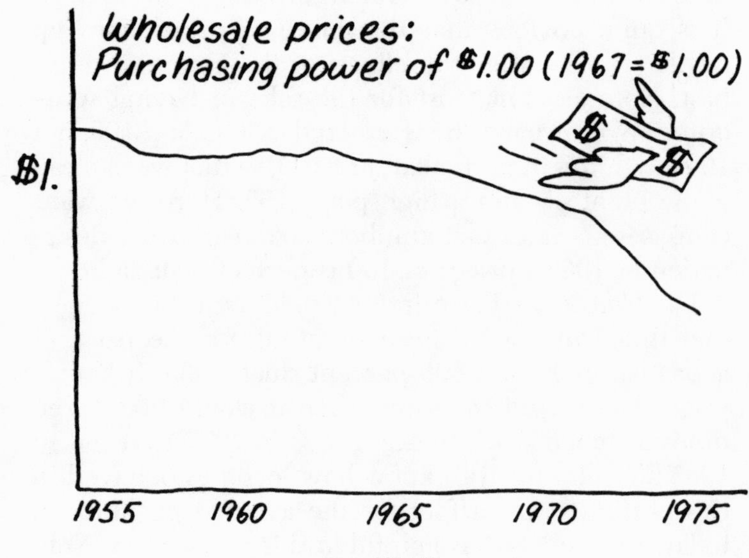

FIG. VI-15

moving picture of the per capita gross national product (GNP). This is not exactly a measure of income, but rather of the total amount of goods and services produced by the country, divided by the entire population. The dotted line, plotted for the same years, shows a parallel curve for average salaries in one profession, public school teaching. It is interesting to note how from 1940 on the per capita gross national product increased at a very steep rate and how salaries, at least in one sector, increased at about the same rate.

Again, amounts are interesting. But even more interesting are patterns of movement, and these can only be compared by drawing line graphs that measure not just quantities but change in quantities.

Our conclusion from these three graphs might be

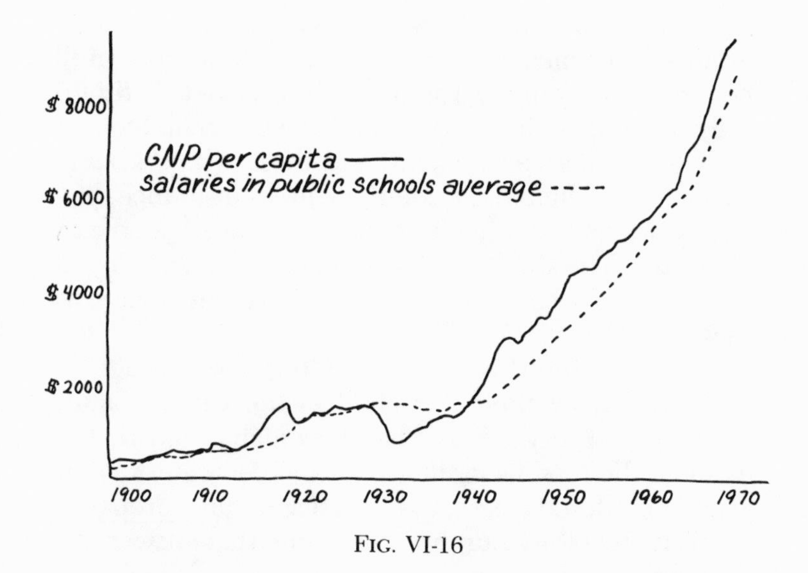

FIG. VI-16

that the American people, on the average, are keeping up with inflation. But what about the others, people on fixed incomes and the unemployed? We might want to look at the unemployment rate over the same period to get a feel for how well some of these other people are doing. Figure VI-17 shows the percentage of the labor force that has been unemployed over the past seventy years. Note the moderate range of variation except for the Depression years. Amazingly, despite the changes in productivity and the increase in the absolute numbers of the population and in the numbers of people, especially women, working outside the home, the unemployment rate has remained almost constant. Critics of the statistical methods used to generate these line graphs of unemployment complain that the way we define the unemployed—as "people actively seeking work" at any one time—causes us to overlook large groups of people: those not able to work because of ill health or inadequate training; the so-called "chronically unemployed;" people who have given up looking for work in discouragement; and those who for some reason are ineligible for unemployment insurance, like farm workers and other seasonally employed people or domestics. The absolute number of people unemployed is probably far larger than the unemployment rate, as shown in Fig. VI-17.

Another interesting statistic is the average age of women bearing their first child in the United States over the past fifty years. What does such a statistic tell us about the reasons for increases and decreases in that age? As is often the case, the statisticians have prepared several interesting numbers, but none that answer our question directly.

Figure VI-18 shows birth rates in terms of age of mother and age of father for one year, 1973. What is

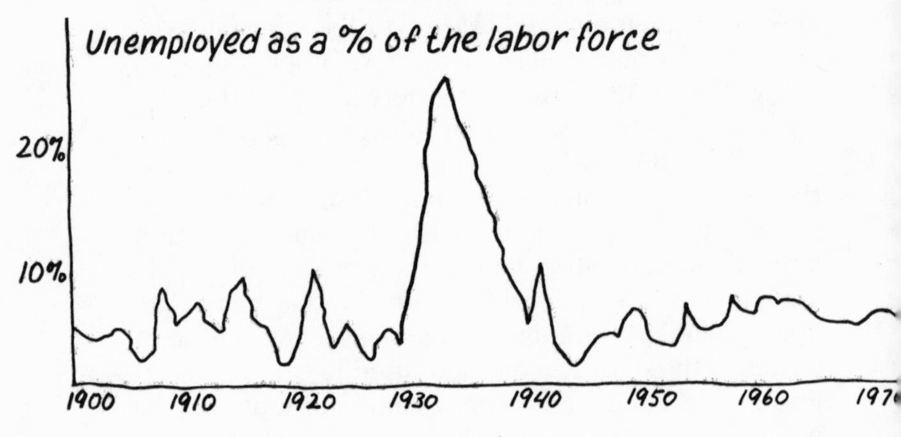

FIG. VI-17

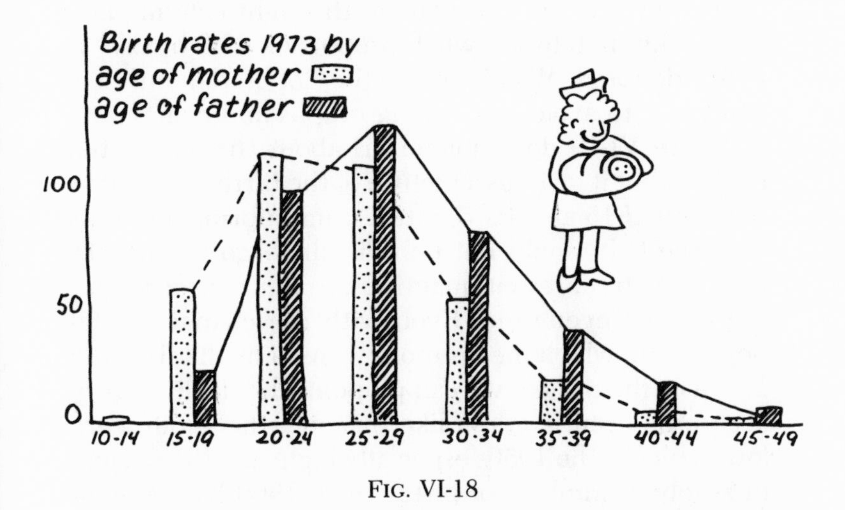

FIG. VI-18

interesting about this display is that at first I thought it represented the numbers of births per age group. If that had been the case, the numbers should all have totaled 1,000 or some other round number. But I found that they totaled less than 350. I was reading the graph wrong (as I often do) and started over. What did those columns mean? They are based on the following: of 1,000 women or men (not 1,000 births) in any age group, how many were having a child in 1973? The answers for women are: less than 10 in the age group 10–14 (which makes sense); about 55 in the age group 15–19; about 115 in the age group 20–24; and so on. This bar graph does not tell us whether these were first children, and so brings us no closer to what we want.

In order to avoid my first mistake, it is important to notice what the graph is talking about, or what is the "of statement"—that group or category that follows the word "of." In this case, the proportions are "of women" in the left column, "of men" in the right column; neither column tells us what proportion "of births" is being denoted. Watching for the "of statement" is a good way to avoid misreading graphs and charts.

Figure VI-19 does not tell us about the age of the mother, but it gives us a picture of the births in millions between 1915 and 1975, interesting information in its own right. It would not be difficult to go from these lines, which represent quantities, to a curve representing change over time. Even with the changing total population, which means more women of childbearing age over the period, we can pinpoint the declines very accurately from a graph like this one: the 1930s were low in births; the 1950s especially high; and the decline in absolute numbers of births since 1960 has been al-

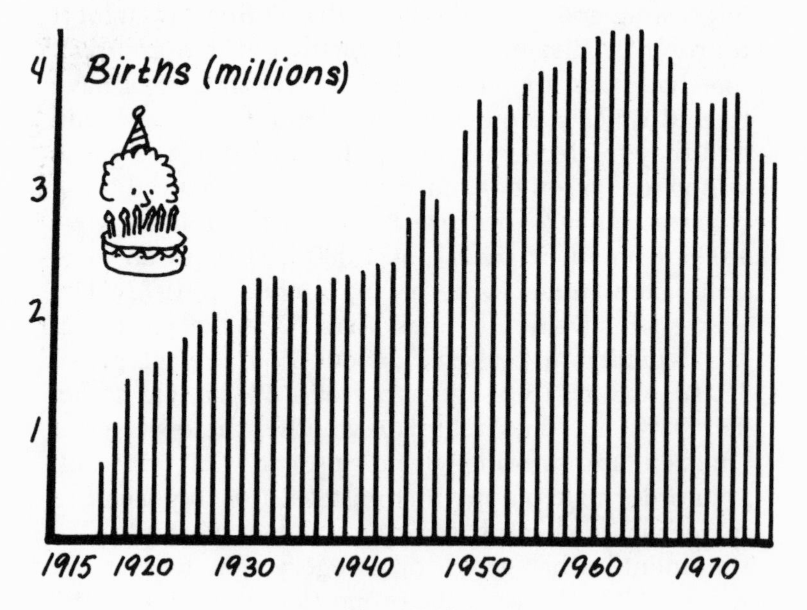

FIG. VI-19

most continuous. Might this be related to rate of marriage? Or of divorce?

Figures VI-20A and VI-20B try to get at these phenomena. The rates are given for 1,000 population;

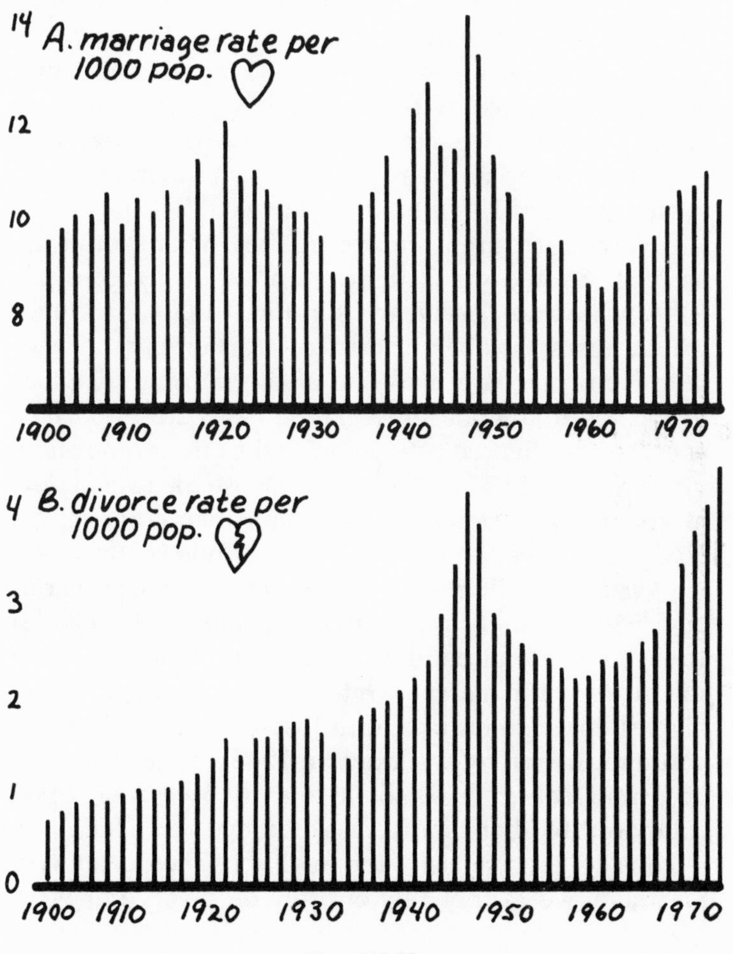

FIG. VI-20

for example, of 1,000 people in 1910, 11 were likely to get married (not *be* married) in that year, whereas during the height of World War II, 16 were likely to get married.* Also, the rate of marriage does seem linked with the number of births because beginning in 1958 or 1959 the number of people getting married (per 1,000 population) was far, far lower than at any other time in this century except for 1931, at the beginning of the Depression. With the number of marriages increasing in 1970, we might (but we also might not) have expected to see an increase in the number of births during the mid-1970s (and we have).

Divorce rates are a stickier matter. What should the base be? Should we compare number of divorces per 1,000 population per year, as in Fig. VI-20B, or should we compare number of divorces to number of weddings? Or to a base made up of the number of intact marriages at any one time? These bases are important because the divorce rate compared to the rate of new marriages might seem high just because the rate of new marriages went down. In 1960 there were about two divorces for every eight weddings, or one in four. This is a very misleading statistic because in fact number of divorces per 1,000 population was down in 1960. In 1948, on the other hand, when wartime weddings were still inflating the marriage rate, the divorce rate was the highest in this century until just recently. But compared to the rate of weddings in 1948, the divorce rate does not look so bad: about four divorces per 1,000 compared to about fourteen weddings per 1,000.

Thus, to get a sense of the quality or the duration of marriage, we cannot rely entirely on divorce or wed-

*These figures might more accurately describe numbers of weddings!

ding rates per 1,000 population. We want to know statistically how many of the marriages made today are likely to end up in divorce. And that statistic is not at all obvious from the tables before us.

The marriage and divorce rates, like the GNP, are very much the product of human choice and human endeavor and hence rather unpredictable. When we take a statistic from the biological realm, such as median height by age (Fig. VI-21), we can see how a moving curve rises and plateaus. The graph in Fig. VI-21 is very interesting in terms of our prejudices about male/female differences. Until age 11, the end of middle childhood, females are either just as tall as males of the same age, or even on the average a little taller. From age 11 on, girls appear to stop growing, at least at the

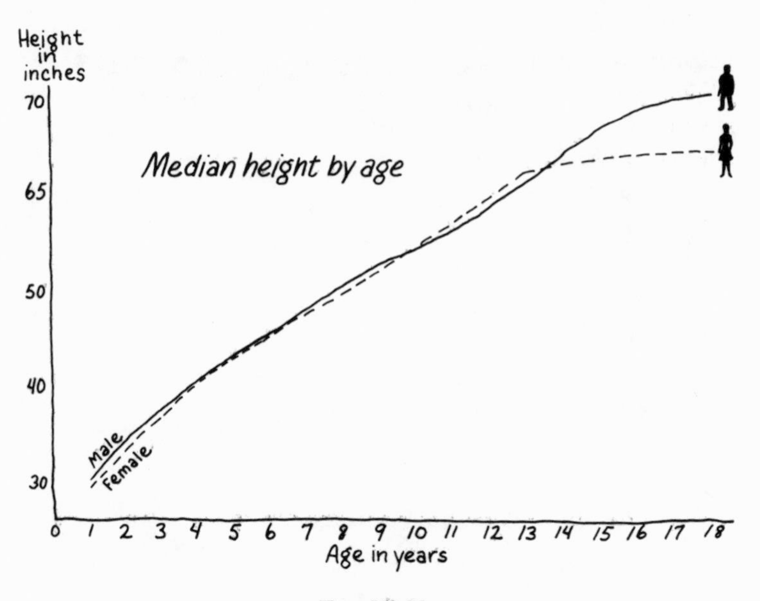

Fig. VI-21

same rate as before, while males continue along generally the same slope. They, too, stop growing, but not until the ages of 16, 17, and 18. Thus, the height difference is clearly the result of a change at puberty and not a pattern apparent since birth.

The significance of such a picture is that other variables confirm the similarity of boys and girls during middle childhood. The differences in height and strength only show up at puberty. Hence, earlier differences in behavior must be social in origin. There is simply no physical reason why an eight-year-old boy should be more interested in and better at sports than a girl the same age.

Fig. VI-22 is a scatter diagram. This one is an attempt to show the clustering of language and science apti-

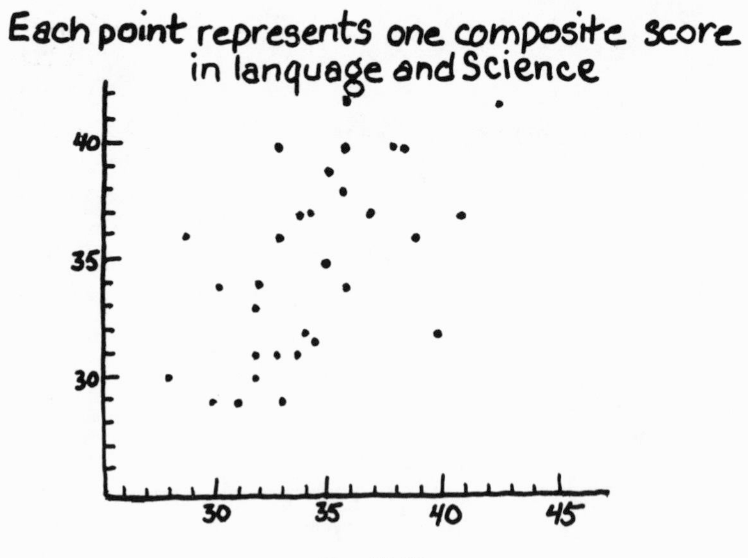

FIG. VI-22

tudes in a certain university population. Scatter diagrams or scattergrams or scatterplots are useful when there are no shapes that can be linked by lines or curves but only clusters of points. Where the points are most dense, of course, one does get some information.

Once one has a line or a series of bars or a cluster of points on a graph, many obvious questions occur that might not have been obvious from the raw data. What accounts for the steepness of the line, for example? Why does steepness increase or decrease at this point? Why is the line straight? Why is it curved?

Comparing data to look for links and associations among trends or events is an obvious next step. Plotting age of childbearing or numbers of children born to mothers of a certain age against some other development, say health of the economy, might give one a clue as to *why.* The advantage of the "moving average" is the dramatic image it gives of what happened over time and under changing circumstances. Everything on a graph can be expressed in words, but it would usually take many words to say as much and the overall impression might not be as distinct.

When we superimpose one moving average on another we are searching for correlation. The figures we have looked at in this chapter provide some good examples of obvious correlation: the GNP-per-capita and the average public school teachers' salaries; the rate of new marriages and the rate of divorce. To be sure that these are true correlations we might want further evidence; also, we cannot know whether one trend causes the other or whether they are in tandem because of a third factor, an "intervening variable," such as the influence of World War II on the high marriage and divorce rates in the 1940s. All too often social scientists who find connections between variables jump to conclusions about the inevitability of their correlation and/or the causative power of one over the other.

"Why" questions can rarely be answered by statistics alone. Why are there more cold days in January than in March? Why has the average age of childbearing for women first gone down and then up? And why has inflation, not a major problem (except during wartime) before 1960, become so severe in recent years? To answer these questions, we need much more than is in the data. We must think up hypotheses, find data describing them, and do more calculations to look for connections between factors. Some forces are quantifiable; many are not. In many cases, careful researchers only conclude that "the data does not disprove my hypothesis." They know that they cannot be sure their hypothesis is true simply by looking at averages and samples of larger populations, however sophisticated their techniques may be.

After one has looked at or drawn many graphs one can get a feeling for the patterns even before one has put pencil to paper. Age and height of growing children is, as we have seen, a picture of upward and forward motion until the plateau is reached at age 11 or 12 for girls and at age 16 or 17 for boys. Being pretty much a straight line, the relationship is called "linear." The relationship between capacity to learn and anxiety, on the other hand, is positive up to a point—one is nervous before a test but should not be paralyzed by it —and then negative. Since this line changes direction, it is called "curvilinear." If the relationships being plotted have two high points or two low ones, like the use of electricity in a large metropolitan area, which will have a small peak in the morning and a larger one between 5 P.M. and 8 P.M., then the curve will be described as "bimodal."

Professional people who look at graphs in their work use a visually oriented vocabulary for discussing issues. The shapes of the curves become themselves a shorthand way of communicating information. "Linear," "continuous," "discontinuous," "curvilinear," "the slope of the curve," and other similar phrases have become a part of common business parlance. There are no numbers involved in these phrases. They refer rather to relationships that have been abstracted from numerical data. And one does not have to be mathematical in order to use them.

At the outset, the newcomer to statistics should decide whether he or she wants or needs a "reading knowledge" of statistics, the ability to get information out of visual or summary statements about data and relationships; or whether the goal is a "performing knowledge," which I take to mean the ability to collect data, select a sample, display information on graphs, and test for reliability or correlations. Few of us will need a "theoretical knowledge" of statistics, an understanding of the theories of probability on which the manipulations are based.

I have learned much about reading graphs by listening to skilled people decode them. People who really need to get something out of this kind of material spend many minutes, sometimes quarter hours, carefully perusing a graphic representation. Like me, they often begin wrong. They miss the "of statement" or overlook the fact that the data is given in thousands, or based on the 1967 value of the dollar. Graphs are harder to read than sentences partly because they are not presented in a linear sequence. One doesn't always start at the left and read to the right, and often the key fact or trend is not even obvious until one has digested the whole picture. Ironically, the process is less like mathematics than like pondering a poem. Poetry, too, is not always understood by reading each line in sequence. A poem should be read many times and, ideally, committed to memory. It helps immensely to discuss it and to read it out loud. A graph, too, in my experience, is better understood when it is shared with someone else. Most important is the commitment we make. My prediction based on a sample of one (myself), is that there is a "positive correlation" between time spent in graph reading and quality of comprehension.

Appendix

A Caveat on Reading Tables

by Allan Johnson, Sociologist, Wesleyan University

Proportions and Percentages

The *proportion* of people in a group who have a characteristic is the number of people who have it divided by the total number of people in the group. The *percentage* who have it is simply the proportion multiplied by 100. If we have 10 people in a group—8 whites and 2 blacks—the *proportion* white is $8/10$ or 0.8; the *percentage* white is $(8/10) \times 100$ or 80 percent.

We use proportions and percentages to make comparisons, either between groups or within groups over time. In reading percentages or proportions, remember:

1. They must all add up to 100 percent unless some explanation is given.
2. If more than one characteristic is involved in a table, be sure to note the direction of the percentaging (across the rows or down the columns). Different directions mean that different bases are being used to compute the percentages and this obviously affects the meaning of the numbers. "Fifty-eight percent of *which group* has *which characteristic?*" This is the kind of question you want to ask.
3. If we percentage across rows, we can make comparisons between columns; if we percentage down columns, we can make comparisons between rows.
4. Before you read a table, study the title to understand exactly what groups and characteristics are included and how they are arranged in the table. Be sure you understand what the numbers in the body of the table mean.
5. Do not take seriously proportions or percentages that have bases of fewer than 30 or so cases.

Means, Medians, Percentiles

For characteristics that can be added, subtracted, multiplied, and divided (such as income, number of children, years in school, age, etc.) we can *summarize* a whole distribution. A *mean* is an average obtained by adding all the scores (incomes, ages, or what-have-you) and dividing that sum by the number of scores you added (number in the group, etc.). A *median* is the score of the middle person when we arrange people in order of increasing income, age, etc. If we have 11 people and order them by income, the income of the sixth person will be the median—half of the people will have lower incomes than the median and half will have higher incomes. A *mode* is the most common score.

A *median* is a better measure of the "typical" case because extreme scores (e.g., millionaires) do not affect its value. The *mean,* however, is sensitive to extremes and is often a distorted measure of the "typical" case. It is very dangerous to compare means and medians, although you will see this done in articles that talk about data from more than one source.

A *percentile* is like a median, but it is more general. A given percentile (say the 60th percentile) is that income, age, etc. that cuts off the lower 60 percent (in this case) of the cases. The median cuts off the lower half (or 50 percent) of the cases and is therefore the "50th percentile." The 90th percentile is that score below which lie 90 percent of all the other scores. Thus if the 90th percentile for income is $15,000, 90 percent of the people make less than $15,000 and 10 percent of the people make $15,000 or more.

Note: If my income is in the 80th percentile and your income is in the 90th, then you make more money than I do. We *cannot* tell from this, however, *how much* lower my income is. We could be quite close in income or we could be quite distant from each other. Thus if one school district is in the 40th percentile and another is in the 70th, this is no reason to conclude that one is much better than the other. On the other hand, if one school is in the 89th percentile and another in the 88th, they *could* be quite different.

Rates and Ratios

A *ratio* is the number of people with one characteristic divided by the number of people with another characteristic. It is a measure of *relative* size. In our earlier example, the ratio of whites to blacks is $8/2$ or 4 to 1, meaning that there are four times as many whites as blacks. Alternatively, the ratio of blacks to whites is $2/8$ or $1/4$, meaning that blacks are only one-quarter as numerous as whites.

Note: ratios measure *relative* size. They do not tell us about *absolute differences.* For example, the phrase "the number of delinquents has doubled" is misleading. An increase from 1 delinquent to 2 delinquents is a ratio of 2 to 1 (a doubling), but it is not very impressive. On the other hand, an increase from 500,000 delinquents to 1 million *is* impressive. Both represent ratios of 2 to 1 but clearly the ratio itself does not tell about the magnitude of the differences or change.

A *rate* is a measure of change over time (mph, population growth, etc.). Many so-called "rates" (homicide, marriage, divorce, etc.) aren't rates at all: they're ratios.

Samples and Surveys

A *population* is any group of people or things that is defined in such a way that we know who belongs and who does not.

Samples are subgroups of a population.

To evaluate a sample we need to know at least the items below. If we do not, we should not take the survey results *too* seriously:

1. Whom does the sample represent?
   What population was sampled?
   Did everyone in the population have a *known* chance of being selected?
   How many people who were selected were not

interviewed? (What was the response rate? Anything under 75–80 percent is in trouble.)

2. How was the information gathered?
   When was the survey conducted? Where? By whom?
   What was the setting in which the information was gathered? Who did the interviewing?
   How were questions worded? What "yardstick" was used?
3. How large was the sample?

You can use the table below to evaluate many of the poll results that use proportions or percentages. For each sample size in the left column, the right-column figure completes the following statement: "We can be 99 percent sure that the percentage (or proportion) in the *population* is the *sample result* plus *or* minus_________."

So, if our sample size is 1,000 and the percentage of our *sample* in favor of proposition A is 35 percent, we can be 99 percent sure that the percentage in favor in the *population* is 35 percent plus *or* minus 4 percent—or somewhere between 31 percent and 39 percent. (If we are talking about proportions, the comparable figures are .35, .31, and .39.)

What follows is valid *only* with properly drawn sample:

Sample Size	We can be 99 percent sure that the percentage in the population is the sample result + or −
50	18 percent
100	13 percent
300	7 percent
500	6 percent

800	5 percent
1,000	4 percent
1,500	3 percent
2,000	3 percent
3,000	2 percent
5,000	2 percent
10,000	1 percent
50,000	½ percent

References

This chapter has benefited very much from the advice on fractions and other arithmetic techniques of Persis Herold, who allowed me to see her forthcoming book, *The Math Teaching Handbook* (published by Selective Educational Equipment, Inc., 3 Bridge St., Newton, Mass.) in manuscript.

On averages and averaging, the advice of Frederic E. Fischer, author of *Fundamentals of Statistical Concepts* (New York, Canfield Press, 1973) and of Joel Schneider of St. Louis has also been helpful.

Two books I enjoyed reading in preparing for this chapter are *Number: the Language of Science* by Tobias Dantzig (New York, The Free Press, 1954); and *Mathematics in the Making* by Lancelot Hogben (New York, Doubleday, 1960).

[1] Persis Herold of the Math Center in Washington and author of *The Math Teaching Handbook* (see above) suggests these definitions and the illustrations that go along with them. George Austin-Martin of Mathematics Education at Stephens College in Columbia, Missouri uses a similar sorting out process with diagrams to introduce fractions to his teachers-in-training.
[2] Herold, *op. cit.* passim.
[3] *The Harvard Magazine,* Summer 1977.
[4] Hogben, *op. cit.,* 178.
Persis Herold derived the graphs from published data.

7

Sunday Math: What Is the Calculus, Anyway?

> The calculus is one of the grandest edifices constructed by mankind.
>
> —Anonymous

Introduction

For those who have not studied it (and even for some who have) the calculus looms large and fierce. Its powers are said to be wondrous, but hard to explain to the uninitiated. The challenge seems to reside in the nature of the subject, its terms, and its fundamental theorems, even more than in its avowed difficulty. But like all great mysteries it does divide us into those who can and those who can't.

For years I had been circling the calculus like a bee attracted by the honey but afraid to land. I had never needed it directly at work, but as a historian I was intrigued that several philosophers had independently but virtually concurrently developed the system as a way to cope with the analysis of different physical problems for which precalculus mathematics provided no formulas, let alone solutions. Occasionally, I would bump up against the language that I gathered was at the heart of the calculus. I could not help noticing that "rate of change," "slopes," and "optimums" were ideas that seemed to enrich my colleagues' capacity to analyze complicated issues. Though I often asked for definitions, something in this subject defied explanation in language I could comprehend. The person I buttonholed would either stutter something about these things being too difficult to explain in words, or rush for pencil and paper to draw some squiggles that left me as befuddled as before.

At times I decided to explore a little on my own and bought books with titles like *Calculus Made Easy* or *How to Have Fun with Calculus.* I made the mistake, however, of trying to leaf through these as I would skim a book on any other new subject, say African art or the history of Byzantium. Even the friendliest preface would soon devolve into formidable examples and esoteric notation. No author seemed able to tell me what the calculus is about without having me learn how to do it. Yet I was stubbornly determined not to learn how to "differentiate" or to "integrate" simply on faith. It was my turn to be rendered impotent by math anxiety. Nothing more, nothing less.

Then, in 1976, a mathematician to whom I had confessed my disability intrigued me with the suggestion that if I thought hard enough, as the medieval monks had, about how many angels could fit on the head of a pin, I might become aware of some of the fundamental issues of the calculus. Angels on the head of a pin, indeed. What could that possibly have to do with slopes, integrals, and limits? Think about it, he said.

And so, with beginning excitement, I went back to look at the problem of the angels and the pinhead, and followed it as I would pursue any line of inquiry, reading, thinking hard, asking questions of others. Slowly the angels helped me evolve a way for me to appreciate the calculus if not yet a way to master it.

For most people, no doubt, the way into the calculus will be through real world problems. How much higher or thicker must a dam be built if the amount of water it is to hold will double in flood years? What is the largest area that can be enclosed within a 100-foot-long fence? What is the distance a ball will travel on the horizontal if thrown into the air at a particular angle and with a particular force? Questions like these and above all consideration of what kinds of methods are needed to attack them will stimulate the kind of inquiry that leads inevitably to the calculus. But for me the monks' foolishness, far more than the real world, opened the door.

Angels on the Head of a Pin*

The monks began with some assumptions. Although we may no longer share their assumptions, they imposed interesting constraints on the problem. God, they believed, was omnipotent. (This was the first constraint.) Therefore, if He wished, He could create angels of any size, small enough, if He so desired, to fit on any piece of matter, even a pinhead. But because God created the universe according to immutable laws of nature (and this was the second constraint), even He could not create angels that took up no room at all or a pinhead that

*I am indebted to Joseph Warren of Mind over Math for this appreciation of the monks' argument. This discussion is, of course, my own.

would contract and expand as the angels climbed on and off. No, the angels and the pin had to be of some finite size. And so the meaning of the problem posed by the scholastics was profound: can an infinite number of finite things fit into or onto a finite space?

Pure logic helps us tackle this question. Imagine any pinhead, that is a pinhead large enough to accommodate two angels of some "normal" size, and imagine our starting with *one* of them in place. Now, God can place as many angels of another size as He wishes in that space so long as the total size of all of them is not greater than twice the size of one of the angels we began with. The most obvious solution is two "normal" angels, but this will not do, for this is a finite number and if God put a finite number of angels on the pin, He would compromise His power. Since He is omnipotent, the number He chooses must be infinite.

If instead He were to start with the one "normal-sized" angel who takes up half the available space and then make the next angel half the size of that one and a third angel half the size of the second one and a fourth half the size of the third, God would never fill the space at all, since one angel plus one-half angel, plus one-quarter angel, plus one-eighth angel, plus one-sixteenth angel . . . plus one one-hundred-twenty-eighth angel added together will never equal two angels. Try it:

$$1 + \tfrac{1}{2} = 1\tfrac{1}{2}$$
$$1\tfrac{1}{2} + \tfrac{1}{4} = 1\tfrac{3}{4}$$
$$1\tfrac{3}{4} + \tfrac{1}{8} = 1\tfrac{7}{8}$$
$$1\tfrac{7}{8} + \tfrac{1}{16} = 1\tfrac{15}{16}$$
$$1\tfrac{15}{16} + \tfrac{1}{32} = 1\tfrac{31}{32}, \text{ Etc} \ldots$$

What God or God's brightest interpreters had come upon was the principle of infinite or unlimited divisibility. God can put an unlimited number of angels on the head of a pin so long as each is half the size of the one that has gone on before. Because the "last angel" can always be divided in half again, the potential number of angels is unlimited. Of course, the interval between each successive angel's size will be smaller and smaller until the difference in angel sizes is almost nonexistent. One may be but one-sixteenth of the original angel in size, and the next may be but one-thirty-second, but however crowded the pinhead, some empty space will always remain. One one-hundred-twenty-eighth is not much room, but it is some.

As to what all this has to do with the calculus, some interesting and important ideas are embedded in the scholastics' problem. I began to examine these one by one.

Space as a Collection of Infinite Angel-Places

Space, motion, and time are difficult to conceptualize, particularly if we try to imagine them having dimensions. To us they are ever in flux. We cannot step into the same river twice because at every moment other millions of droplets of water are moving through and changing the nature of the river. Except with a still camera, we cannot stop a bird in midflight. A circle seems to have no beginning or middle or end. Yet, if we are to measure space, time, and motion, if we are to express these phenomena in numbers, we must find a way to count them, to divide them into parts we can handle.

Facing this dilemma, Greek mathematicians very

creatively defined time as a collection of instants and space as a collection of points. The idea was to have something concrete to add up. This of course led them into deep paradoxes. If time is a collection of discrete instants, then an arrow flying through space could be said to be never really "in motion" at all, for at every single instant in time it would be at one and only one point in space. Therefore, to be logically consistent, one would have to describe it as being at rest.

These paradoxes notwithstanding, the conception of space and time as divided into infinitely small, closely packed points or instants fits our intuitive notion that time and space are in flux; for if the points can be made close enough they will blur like raindrops into something that seems to be not drops at all but flowing water. Or, in another writer's analogy, we know music is made up of individual sounds, as in a chord or a melody, but when we listen the many sounds seem to make but one.

The power of this idea is that if we can divide space and time into points and instants, we can begin to relate something happening in one of these two dimensions to its passing through the other. And this possibility gave ancient mechanics and modern physics a way to begin to measure motion and force.

Diminishing Differences

If the first two angels in the series described above are one-half "normal" angel apart in size and the "last two" are only tiny fractions of the pinhead apart in size, then we can conclude that the very last angel (if there is such a one) will be practically indistinguishable in size from the one that preceded her. The fact that the more angels we invent the smaller and less significant are the differences between them will provide a very helpful procedure in approximating things we cannot exactly measure or count.

Unlike the mathematics that precedes it, calculus deals with things that often cannot be precisely ascertained. The position of an anchored ship vis-à-vis a lighthouse or the unchanging area of a rectangular plot of land are not calculus problems. Calculus deals with dynamics. One way to view the calculus is as a system for getting a mathematical grip on events that don't stand still long enough to be measured.

To "get at" an event as it is happening is not easy. One can get a sense of the situation "before" and "after," but how do we put "during" into a box? "During" can be roughly considered as the space or time or distance or gap between "before" and "after." In fact, the nearer "before" is to "after," the more precisely we can measure "during." Any moment or half-moment or quarter-moment or one-hundred-twenty-eighth of a moment before something happens is still "before;" any moment or half-moment or quarter-moment or one-hundred-twenty-eighth of a moment after something happens is "after." The closer we can get to the "during" by diminishing the difference between "before" and "after," the more likely we are to capture it.

This is why it becomes useful to measure distance or difference between conditions before and after an event. The smaller the difference between the before and after phase in time, or the in-front-of and the just-behind phase in space, the closer one can come to describing the event, the change, the spot, or the speed itself. Think of movies, which are fast-moving still pictures flipping over fast enough to fool our brains into behaving as if the pictures are in flux; or animation, based on the same principle. The idea is not so different from squeezing angels onto the head of a pin. At no point will the difference between before and after, or the addition of the next smallest angel, be nothing at all. But the closer that difference is to zero, the closer one has come to the event (the adding of the "last angel") itself.

The trick is to imagine an infinitely small interval, a "curve" of zero length, a "line segment" that is barely more than a point, a tiny "difference" so small that it is like no difference at all; and to place this imaginary construct upon the unmeasurable or uncountable event.

Functions

It is because calculus deals with "events" that it is so hard to grasp. Consider a key concept of the calculus described as a "change in one variable with respect to a change in another." Until now, we have associated mathematics with counting and measuring. But here we encounter not entities but relationships or functions, which are defined in terms of one or more others.

In the case of the angels, the function is the relationship between the number of angels and their size. That function can be plotted on a graph to look like Fig. VII-1. From this graph we can appreciate that the size of the individual angels will never be zero and that the total area taken up by all the angels will approach a finite amount.* There is a relationship between the size of the angels and their number. As the number of angels changes, their sizes change, when number is 1, size is 1; when number is 2, size is ½; when number is 3, size is ¼; when number is 4, size is ⅛, etc.

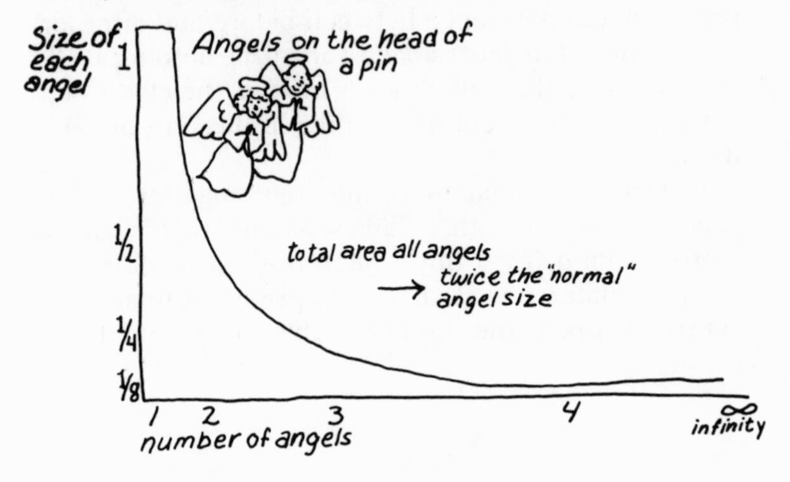

FIG. VII-1

In the same way, though depending on many more variables, each with its own pattern or effect, a growing plant develops at a rate that results from interaction among all of the following: the species of plant, the amount of light, the stress factors affecting the plant, the amount of fertilizer, and the size of the leaf area. The addition of plant fertilizer (or the change in one entity with respect to the growth rate of the plant) is not going to be a fixed quantity at all. It is more likely to be a function of how much fertilizer has already been

*For the sake of the example, we are imagining one-dimensional angels, having only height not "size" in the ordinary sense of the term.

used and how much light, water, and other factors will continue to be available. Some amount of fertilizer may help the plant and the very next increment might damage or even kill it.

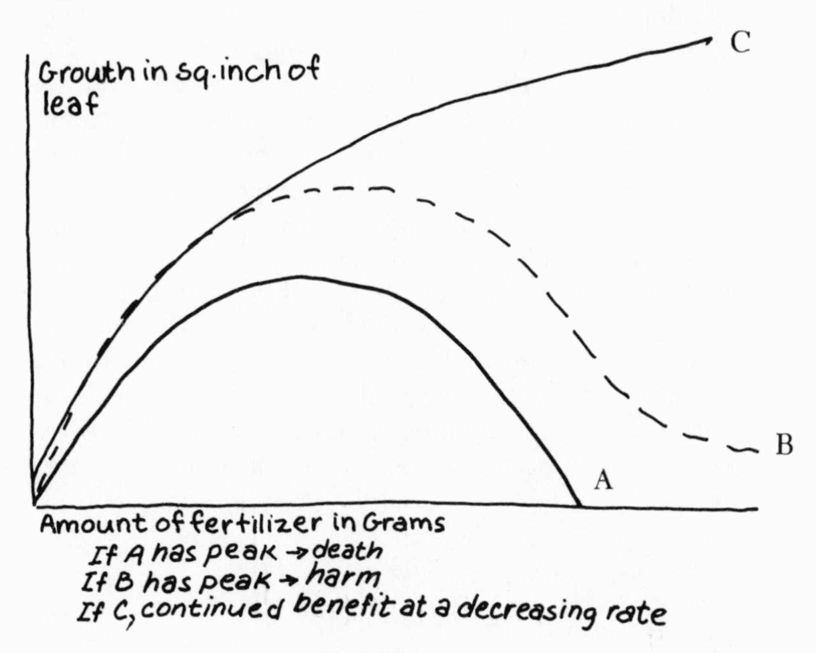

FIG. VII-2

Figure VII-2 shows three possibilities: A. that the addition of plant fertilizer peaks very soon and too much destroys the plant; B. that the benefit of additional fertilizer peaks fairly soon and too much only harms the plant; and C. that the addition of fertilizer continues to be beneficial though at a decreasing rate.

Bread baking is a function of so much yeast, reacting in so much dough. How do we arrive at satisfactory amounts for each of these elements and what will happen to the kind of loaf we bake if we change one or more of these in one or more directions? Just as the relation of plant fertilizer to plant growth will not be a constant, so the effect of adding more yeast to the bread will depend, among other things, on how far into the rising process the bread has already gone.

In the case of the angels, as we let one variable (the number of angels) grow, the other variable (the size of each angel) got very small and the third variable (the total size of all the angels together) approached the size we called two "normal" angels. This is a special kind of relationship: the series is called convergent because the total size of the angels approaches a finite number.* But it is typical of the calculus dynamic in that the change in one variable may affect the change and the rate of change in the other two in different ways. In mathematical terms, the tending of the littlest angel to zero size and the tending of the total quantity to "two" are "limits."

The Idea of a Limit

As we have seen, the number of angels that can be fitted on the head of a pin can be infinite even though the total amount of space they will occupy is finite, so long as the size of each angel is successively half as large as the one created before. No angel will be zero in size. But the "littlest" angel will be small enough to be very close to zero. Our problem is correctly formulated when we figure out that zero is the limit of one variable, two the limit of the second,

*A series might be divergent or even oscillating, instead.

and infinity the limit of the third.

In cases where we cannot firmly grasp a variable such as instantaneous change we are going to have to conceptualize our solution by creating ever smaller intervals that approach the exact moment of change. When we talk about measuring the "slope of a curve," for example, I find it useful to imagine a line of some length in the curve which closes in on the curve getting shorter all the time. At the point where it will actually touch the curve it is a "line of zero length." Since there can of course be no line of literally zero length (just as there can be no angel of zero size), we shall start with a line of some length and approach a line of virtually zero length, presuming zero length to be the "limit" of our progression. (See Figure VII-3.)

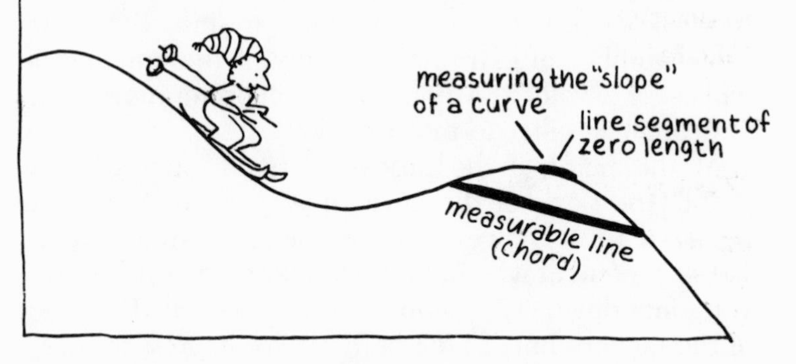

FIG. VII-3

Mathematicians conceptualize this a little differently. They imagine a line cutting across the curve too; but instead of moving the line forward toward the curve as I do, they rotate the line until it becomes congruent with the curve. For me rotation is harder to imagine than moving forward and so long as I can cope with the idea of a line of zero length I feel comfortable about thinking about the problem my way.

Maximums and Minimums

"Limits" can be helpful in another sense. In real life we may want to know at what point a procedure or policy or manufacturing plan reaches its optimum usefulness; or, to say it differently, the moment just before there are diminishing returns. If we have a system for measuring continuous and instantaneous change, we can find out pretty easily at which point in the process *no change* is taking place. For an increase in pay per hour, for example, some number of employees will be willing to work extra hours or an unpopular shift. But after some number of extra hours, many of the employees will cease to value the money and will prefer more time with their families to more work.

At this point, if the process is being plotted on a graph, the slope of the curve changes direction. Just before it turns, however, it reaches a maximum (Fig. VII-4), a point at which it is no longer going up, but not yet going down. The "slope of the curve" or the slope of the tangent line, as mathematicians express it, measures zero at this point. And this is precisely the point we want to be able to locate to find the optimum arrangement to keep worker incentive high.

To take another example, some cutback in expenses in running a store will not affect revenue. But at some point, such as closing the store on Saturdays or removing the neon sign, a change in policy will discourage so

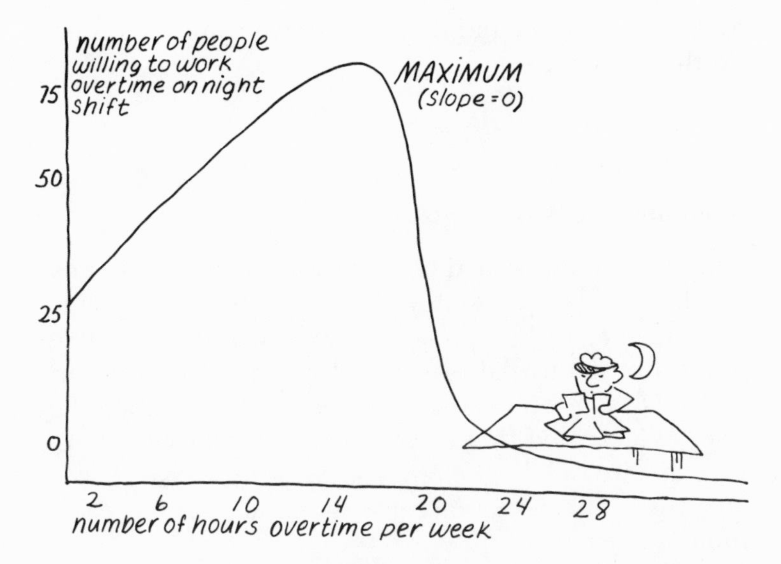

FIG. VII-4

many customers that there will be a change in the direction of gross sales. In fact, this pattern is very common in certain real-world situations. One hits a peak, after which there is a flattening out, a plateauing of the effect. This, too, can be ascertained mathematically through techniques that trace change as it occurs. If we know the "slope of a curve" and its high and low points, then we have valuable information about a system: its rate of variability (the pace at which it varies or changes) and the limits to that variability (the conditions at which the change ceases to take place).

Thus after making a mathematical model of the situation, the calculus permits us to find maximums and minimums, points at which the "slope" is zero, and this can answer many real-world questions. Often we do not have to be able to do these calculations ourselves. So long as we understand their power we can have others do them for us.

Summing the Angels' Sizes

Suppose our angels had been decreasing in size, not by one-half (1, $\frac{1}{2}$, $\frac{1}{4}$, $\frac{1}{8}$, $\frac{1}{16}$, etc.), but in the following sequence: 1, $\frac{1}{2}$, $\frac{1}{3}$, $\frac{1}{4}$, $\frac{1}{5}$, $\frac{1}{6}$, etc. How might we have found out if this pattern would result in the convergence of their total size toward some finite number like two? We can see that their individual sizes would approach zero but $1 + \frac{1}{2} + \frac{1}{3} + \frac{1}{4} + \frac{1}{5} + \frac{1}{6} \ldots$ etc., as it turns out, does not converge at all, but increases infinitely.

Instead of working this out arithmetically, we might have found out whether this relationship had such a limit by another process. Mathematicians very often like to solve a problem geometrically. Drawing the function on a graph (see Fig. VII-1, again) and finding the area under the curve would have given us our answer.

But how does one find the area under a curve? Circles are one thing to measure. They are closed and regular. It is quite different to measure area when the shape is so nonuniform. One method worked out two thousand years ago imposes an infinite number of measurable rectangles (Figs. VII-5 and VII-6) until the total area of the rectangles begins to approach the area under the curve.

Compare Fig. VII-5 to Fig. VII-6. The curve in each case is identical, but the rectangles that have been drawn within it are half as wide in Fig. VII-6 as in

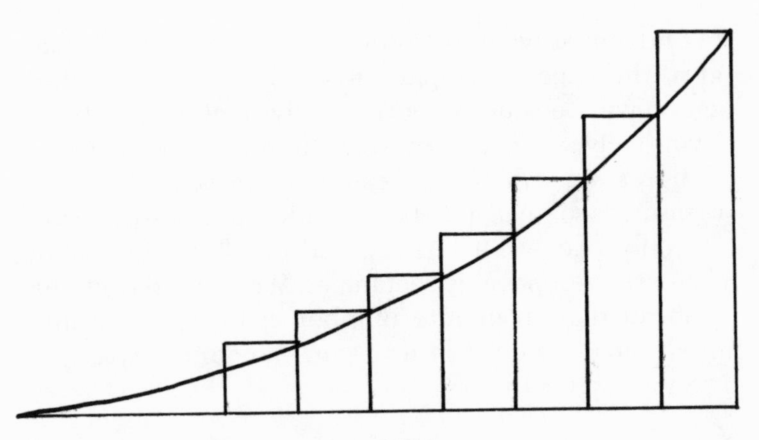

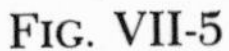

FIG. VII-5

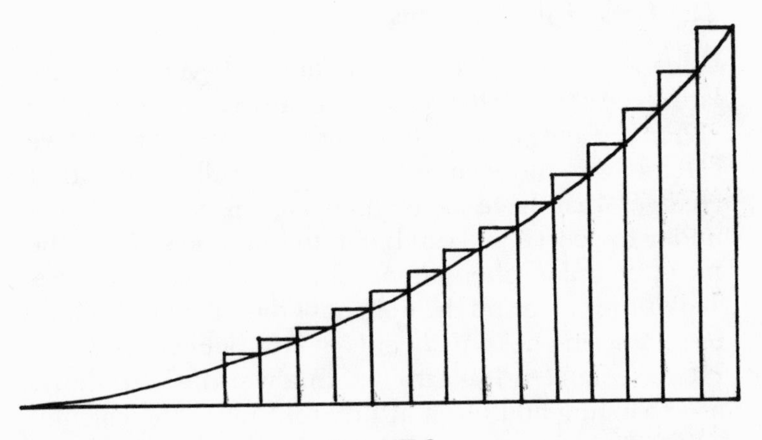

FIG. VII-6

Fig. VII-5. Here the artist is trying to create something that can be measured, place it on the unmeasurable space, and then reduce in size the little triangular chips that extend above it. If we could add up all the rectangles the area would be larger than the one we are trying to find; but as we draw more and more narrow rectangles, the excess (compare Fig. VII-5 and Fig. VII-6 again) will become proportionately smaller each time. Eventually, we could say that the difference between the area we *can* measure and the area we *want* to measure is so small as to be insignificant. But we can say this only when we have imagined an infinite number of the narrowest possible rectangles. We learned from the pinhead that an infinite number of things of an infinitely small size can fit into some amount of space.

So much for the angels.

The Uses of the Calculus

Once we can capture an instantaneous event by calculating the change that took place between "before" and "after" by finding the "slope of the curve," and once we can measure the accumulated effect of all the events or changes that have taken place by finding the area under the curve, we can begin to examine some of the more complex phenomena of the physical universe. This is the value and the delight of the calculus: it allows us to describe not only events in flux, but events whose rate of change is itself in flux. In a world where things are changing and often at different rates, the calculus is indispensable.

Let us look more closely at how the calculus does this. We will begin by graphing the motion of an object moving at a constant rate, not speeding up or slowing down. Now if an object is not speeding up or slowing down we can think of it geometrically as regularly covering the same amount of space in the same amount of time. Therefore, the most meaningful and precise way to describe the object's rate of movement (how fast or how slow) is in terms of the distance it travels over time. We all do this every time we talk about automobile speed. The rate is always so many miles (distance) per hour (time). Using these two dimensions, we can graph the motion of that car at two different speeds (Fig. VII-7). The two speeds, 50 mph and 75 mph, have different steepnesses, but both are still represented as straight lines because in each case a certain amount of space is covered in each succeeding interval of time. The straight-line part of this picture tells us just that: the speed in both cases is uniform.

In plotting the flight of an arrow, however, we are dealing with something much more complex. For one

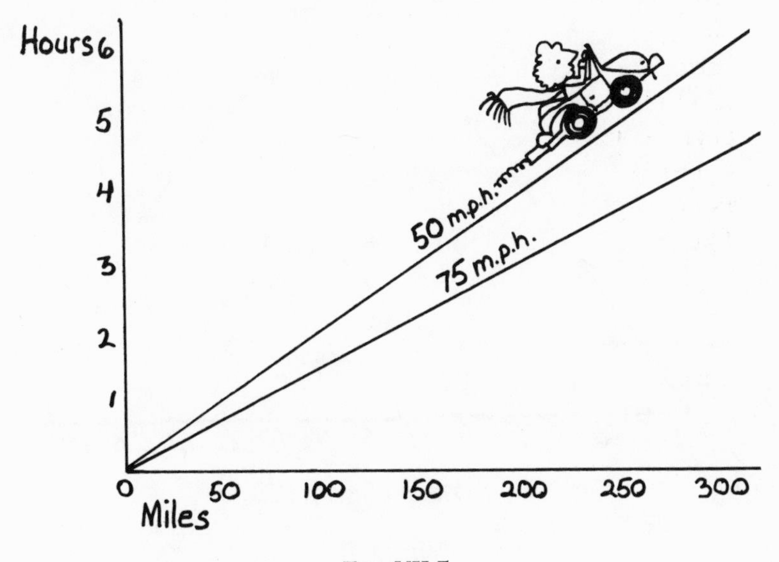

FIG. VII-7

thing, although the arrow can be said to be going at one speed at any given instant and at another speed for another given instant (as the initial momentum weakens and the force of gravity acts on the arrow), the change will not be sudden or abrupt (unless the arrow hits something in flight) but gradual and continuous.

Moreover, the arrow's "speed" has to be considered in two ways: the speed with which it is going up or down and the speed or distance over time at which it is going toward its destination. Consider the different

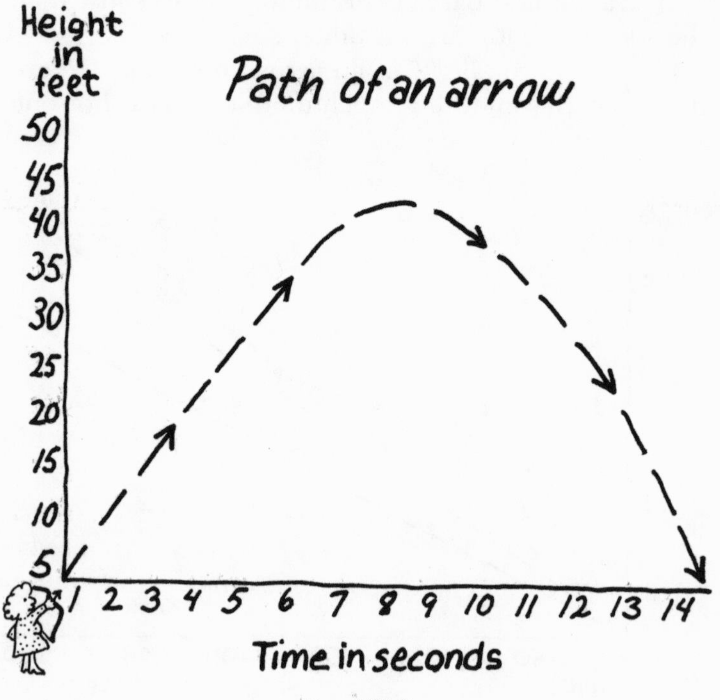

FIG. VII-8

paths described by a fast ball pitched from the mound to the batter and a high fly batted into the air. "Speed" is no longer something that can be talked about in simple terms. In fact, while the path of the arrow can be graphed to look like a rather regular curve (Fig. VII-8), the path of its speed, measured at every instant, is going to look like Fig. VII-9.

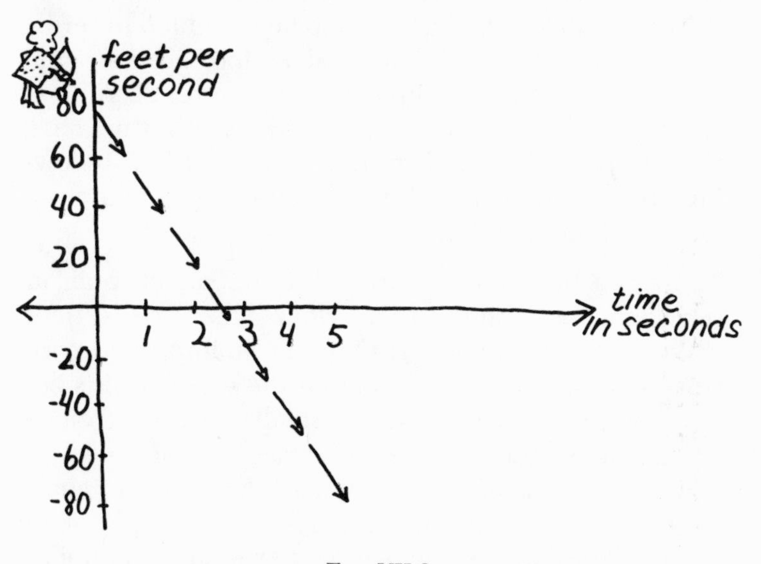

FIG. VII-9

Now, if we can measure how the slope changes in moving along the curve in Fig. VII-9 we can get a sense of how much the arrow is accelerating, decelerating, and then accelerating again. Then, if we can get a sense of those various speeds as effects of various forces on the arrow, some kind of generalized pattern may emerge to predict once and for all what will happen to a body of a certain shape when thrown with a certain force:

how long it will fly, at what speed(s) it will travel, its rate of acceleration and/or deceleration, when and how it will change direction, when it is likely to land, and so on. Answers to such questions, if they are generalizable, are useful in predicting the decay of a pendulum swing, the trajectory of a bullet, or the flight of a rocket through space. And all these generalizations emerge from a consideration of changing rates of change.

Not all rates of change have to be imagined in terms of motion and speed. Dimensions, too, are variable. Engineers use the calculus to try out different combinations until they arrive at the length, width, thickness, height, span, and whatever else they need. The following problem appears static in that it involves an immobile building. But it is also a calculus problem:

There is a building in Hartford, Connecticut, built in the shape of a ship to "overlook" the Connecticut River. It is quite obvious that the building was conceived of and constructed before the energy crisis because it provides a relatively small volume of office space inside a relatively large surface area of walls and windows. For a lot of wall space, there is minimum usable space, leading to high cost and low payoff in rentals. Suppose one had wanted to reshape that building during the design stage in order to increase its volume, using the same amount of steel and glass? How could one have created maximum volume in a minimum envelope?

The solution to this problem will involve calculus because area and volume (like angel sizes and number of angels) do not change in a constant relationship. Yet a relationship does exist, and, depending on the structure, it can be expressed as a formula and plotted on a

The Phoenix Building in Hartford

graph just as speed can be graphed as a ratio of distance and time. Indeed, so long as *any* force can be understood as changing in relation to something else (heat or pressure, size or mass, amount of oxygen, or whatever) and so long as the precise nature of the relationship of these changes can be formulated, these forces, like time, space, and motion, can be captured, tamed, and counted up. From this perspective, for example, the rate at which heat is transferred from one element to another depends upon a variety of conditions that can be isolated and measured: what the thing is made of, how cold it was to start with, the elapsed time, and so on.

Bring in a thermometer from outdoors one day and note the temperature registered on it.* Note also the temperature of the room and then monitor on graph paper the rate at which the thermometer warms up. If you do a careful job, the graph should look something like the one in Fig. VII-10. From the observations alone we can plot a curve to represent the rate at which the thermometer warms up over time and even measure the area under the curve, the accumulated warming up. What is missing from our understanding of the phenomenon of heat transfer is the equation or function that expresses the relationships among the variables—mercury (in the thermometer), heat differential between the room and the thermometer at the beginning of the experiment, and elapsed time. We can see this relationship in the curve; now we have to go backwards from the curve to find the function that produced the curve. (I am grateful to Eli S. Pine, author of *How to Enjoy Calculus,* for permission to use this example.)

First the temperature rose rather rapidly and then it seemed to taper off. But at what rate was it rising, was the rate constant or continuously changing, and how

*This example and the graph are borrowed from *How to Enjoy Calculus,* written by Eli Pine, Copyright 1975, and published by S.H.C. Publishing Co., Box 187, Hasbrouck Heights, New Jersey 07604.

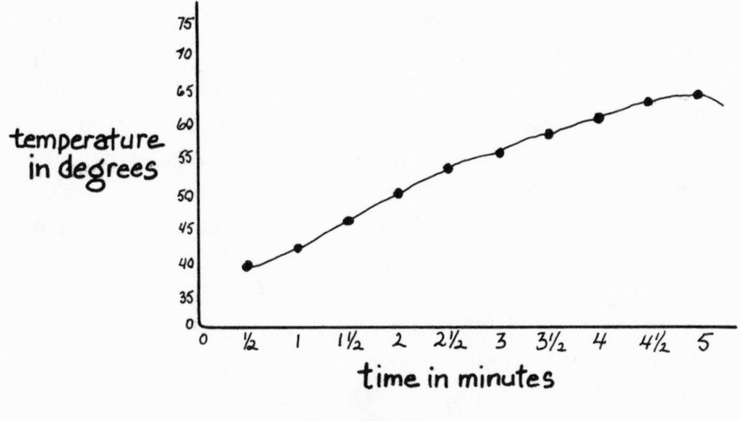

FIG. VII-10

long and in what fashion will it plateau? Using calculus, a mathematician can derive the function that reproduces the rate at which the thermometer warms up. The mathematical relationship does not give us any explanation for the process. Rather it provides a kind of mapping of the effect, enough to permit us to look more carefully at the variables, to predict what will happen next, and to begin to ask the "why" questions more precisely. This is the contribution of mathematics to science.

Conclusion

Since change may take place in such diverse variables as heat and cold, size and speed, growth and decay, all these phenomena will be amenable to mathematical formulations. Pressure and combustion quite as much as speed and time can be expressed as one variable changing in relation to others. Just as number permits us to count everything from objects to abstractions, mastery of the calculus permits us to grasp relationships between and among things in flux. Indeed, when Newton first developed the calculus, he called it "fluxions," which well communicates its essential power: to cope with and measure the process of change.

Where does one go from here? If the calculus were only a method for solving engineering problems, I would not go any further. But it is far more. It is a way to think about things that suggests relationships among them and a way of ranking their impacts. And, since it offers a way to cope with independent and interdependent things happening at the same time (such as the value of the dollar declining overseas, exports going up but imports becoming more expensive), a person trained in the calculus will not have to oversimplify a problem to solve it.

This is the good news: out there just over the horizon for many of us is a mathematical system that can enhance our ability to think and increase our mastery of complex issues. Now for the bad news: there is probably no short cut to mastery. We can talk about the calculus only so long. At some point, as when we learn a foreign language, we will have to memorize the vocabulary, master the rules of grammar, do the exercises, mimic the movements of the experts, and solve actual problems.

Getting from here to there seems easier now, at least for me. By the time you read this, I will have overcome my resistance to "differentiation" and "integration" and taken my first course in calculus. For me, reflection on the meaning of infinity—what infinitely many slices of infinitely small portions will do—set me thinking about divisibility, intervals, limits, and sums as I had never thought about them before. As my awe of the calculus increased, my fear of it decreased, and when I substituted words that made sense to me for the alien notation, I began to integrate the calculus into my own way of thinking about things. Eventually, the subject became a collection of ideas, not just another branch of mathematics.

"Infinity" need not be the starting point. Think about "speed." How would we explain to a Martian who had never seen a speedometer before what our "speed" in a car actually is at any one instant? If we could not use distance and time as we do when we call our rate of travel "so many miles per hour," how would we explain what "speed" is? Or think about the rowboat going upstream at six miles an hour against a current going downstream at three miles an hour. What is occurring at any one moment as the two forces (your rowing and the stream's current) act jointly upon the boat? Starting from questions like these or from a consideration of "decreasing increases" or "increasing decreases," we might end up with kinds of issues similar to the ones that led me into the calculus.

Somewhere out there for each of us, I believe, is a metaphor that will make sensible the shapes and squiggles that give us pause when we open a calculus textbook. And to open our minds to the possibilities of the calculus, any metaphor that works will do.

References

This chapter has been aided enormously by conversations with my friends and colleagues who really understand the calculus. Susan Auslander, Joel Schneider and Peter Hilton were particularly helpful in steering me away from pitfalls and encouraging me to pursue my own line of thinking. Persis Herold and Susan Auslander designed the graphs. Eli Pine's *How to Enjoy Calculus* (Hasbrouck Heights, N.J., H.S.C. Publishing Co., 1975) provides a kind of overview that the uninitiated can read without at first doing any of the problems. Someday, though, somebody should write a calculus textbook for people like myself.

8

Overcoming Math Anxiety

> The best therapy for emotional blocks to math is the realization that the human race took centuries or millennia to see through the mist of difficulties and paradoxes which instructors now invite us to solve in a few minutes.
>
> —Lancelot Hogben, *History of Mathematics*

Quite probably, hidden in our minds in recesses that are not easy to expose, is much of the math we need to enhance our occupational choices and enrich our personal lives. In digging out these dimly remembered procedures we may notice some of the mathematical ideas that escaped us the first time around (as we saw in Chapter Six) and be ready for more advanced and stimulating math (like the calculus). To get at these recesses, I believe, we need only some knowledge about our own minds and how they work, about our temperament and fear and how these intrude on good thinking, and some simple exercising of our rusty pipes. If we can control our anxiety by recognizing its symptoms and coping with them, we can go far.

How far do we want to go? Not far enough to become

engineers or mathematicians, perhaps. But surely far enough so that our fear of math no longer makes decisions for us.

What follows is no surefire cure for math anxiety, but a set of suggestions. Some of these techniques involve professionals. Math teachers or counsellors or both may be necessary at the outset. Other techniques can be developed by oneself and for oneself. Some concrete materials, readily and cheaply available, will help elucidate certain numerical and geometrical relationships. Still other concepts may require some reading in the history of mathematics or in the psychology of learning to provide some theoretical underpinning for what we are doing. All require that we take some initiative in our learning process. As W. W. Sawyer, a mathematician and an inspired teacher of mathematics, puts it:

> In discovering something for ourselves, we have a sense of freedom and conquest. In memorizing something that another person tells us and that we do not understand, we are slaves.

Math Anxiety Bill of Rights
By Sandra L. Davis

I have the right to learn at my own pace and not feel put down or stupid if I'm slower than someone else.
I have the right to ask whatever questions I have.
I have the right to need extra help.
I have the right to ask a teacher or TA for help.
I have the right to say I don't understand.
I have the right not to understand.
I have the right to feel good about myself regardless of my abilities in math.
I have the right not to base my self-worth on my math skills.
I have the right to view myself as capable of learning math.

I have the right to evaluate my math instructors and how they teach.
I have the right to relax.
I have the right to be treated as a competent adult.
I have the right to dislike math.
I have the right to define success in my own terms.

Getting Help

Group Desensitization

A group of women undergraduates at a midwestern college were gathered in a comfortable living room. The two group leaders, a mathematics professor and a counselling psychologist, had worked with them twice before. In the first session, the leaders had explained who they were and what they wanted to accomplish. Then they had gone around the room eliciting math autobiographies. "Tell me about one negative experience you had in learning math," the psychologist asked. Out came all the pain and struggle, much of it from the elementary school years. "Now tell me if you can recall any positive experiences." There were those, too. Curiously, some of the negative memories included positive experiences and some of the events remembered positively were negative at the same time. Ambiguity, confusion, and defeat were the themes of these recollections.

During the week preceding the third session, the one I was to observe, the students had been asked to keep a math diary. What were their feelings when they had to use math? Had they been able to approach math a little more enthusiastically than before? Above all, they were to make conscious all the little things they were doing to themselves when confronting math. Ellen began. "I balanced my checkbook this week," she said with pride. How did she feel when she was doing it? "Well, I noticed that I trust my calculator more than I trust myself, even when the addition and subtraction I do come out right."

Susan's checkbook had not balanced. She had been a few dollars off and decided to wait until her boyfriend came to visit on the weekend to have him find the error. It was embarrassing for her, not just because she'd had to quit but because "I do my math so childishly," she explained, marking down all the carry-overs. Because of this she hated to have her boyfriend see her work. But she showed it to him and defended her method and he found the error. The psychologist commented that some people white-ink their previous additions and subtractions when they find a mistake so that they can start over afresh. It is hard to find a mistake, though, when you keep going over the same computations again and again. Everyone nodded in agreement.

Did you ever try doing digital summing, the visitor asked. No, what's that? Well, it's a trick to check to see whether your addition or subtraction or even multiplication is correct. You add up the digits sideways, until you end up with a one-digit number. If the addition or subtraction problem was done correctly, then the digital sums will add or subtract right, too. Show us.

The math teacher went to a flip chart and wrote:

$$
\begin{array}{rlllr}
\$34.67 & \to 3+4+6+7 & \to 20 \to 2+0 \to & & 2 \\
\underline{-2.89} & \to 2+8+9 & \to 19 \to 1+9 \to & 10 \to 1+0 \to & \underline{-1} \\
\$31.78 & \to 3+1+7+8 & \to 19 \to 1+9 \to & 10 \to 1+0 \to & 1
\end{array}
$$

Checks out. Oh, wow, said two of the young women. Marilyn said nothing. Have you noticed, the math teacher went on, warming to his subject, that you could have eliminated all the digits that add up to 9 when you do digital summing and ended up with the same digital sum? $19 \to 1+9 \to 10 \to 1+0 \to 1$. Eliminate the 9 from the 19 in the first place and you have the 1 you are going to end up with anyway. This is sometimes called "casting out nines" and it makes the process go even faster. Oh, wow, said two of the girls. Marilyn said nothing.

Now, do you know why we can cast out nines? It has to do with our number system being based on ten. The math teacher was getting carried away, so the psychologist intervened. Just a minute, now. Did all of you follow what has been going on? Do you understand what digital summing is all about? Several nods. What about you, Marilyn? I figured I could not learn it. It was too unfamiliar. So I would just do my adding and subtracting the same old way. Is it because you don't trust it really? Because in the past people have told you things were easy and they turned out not to be easy at all? Is that the reason? Maybe.

The math teacher went on. This business of casting out nines can help you balance your checkbook in another way. Say you are off by 45¢ or by $5.40 or some other multiple of nine. Chances are that you (or the bank) have reversed some digits in the computation. You see, when you misplace digits in a number, the following kind of thing happens.

On the board:

$37.18	(the bank's balance)
−31.78	(your balance)
$ 5.40	(9 × 60¢)

Try it another way. Suppose you carelessly transpose \$37.18 into \$18.37. Subtract \$18.37 from \$37.18 and you get:

$$\begin{array}{rl} \$37.18 & \text{(the bank's balance)} \\ \underline{-18.37} & \text{(your balance)} \\ \$18.81 & (9 \times \$2.09) \end{array}$$

It always works. Therefore, when you find your checkbook is out of balance, divide the out of balance amount by nine. If it divides evenly, then check your work for transposition of digits.

I asked the math teacher later if he had intended to get into digital summing or to mention the magic of nines in this section. No. Nothing is planned beyond the assignment and one exercise. Yet the experience seemed to have coherence for us. We started out talking about balancing checkbooks and we ended up with a very useful trick to help us find mistakes. This is one of the differences between a math anxiety workshop, such as this one, and a math class. The group moves at its own pace. The counsellor keeps the math teacher from pursuing more math than everyone is ready for, and people enjoy what happens because they feel a part of it.

The conversation went on. Angela had had an interesting experience with math that week. She had been reading a business textbook and found that she could understand the equations quite well as they were pictured, but not the verbal explanations. She would go over the words again and again, feeling adrift, and would have to go back to the equation each time to make sense of it again. Is it because the words are not the kind of words you would use to make sense of the equation, the counsellor asked. Did you ever think of trying to write out the explanation as you would best understand it and then comparing your paragraph with the one in the book? This might help you see which kind of words and sequences of statements make sense to you and which ones don't. No, she had not thought of that, but she would try it. Perhaps your paragraph, Angela, would have more meaning for me than the one in the book, someone else said. I'm just the opposite, said Ellen. I'm a language major and I tend to skip over the formulas and the equations and rush for the words. Only words make sense to me. But sometimes the words in a math book don't and then I feel really abandoned because there is nothing I can do about it. At least you can understand the equations, Angela. Yes, Angela thought she understood the equations quite well, but even so her problem with the words made her lose self-confidence.

Paula had had another kind of experience. She had divided up the phone bill among her roommates this week and had learned to mix the chemicals in the photo lab where she works. Until this week, she had relied on someone else to mix them for her. As she told of her accomplishments, she looked very pleased with herself, almost smug. Not as hard as she had expected. Did you feel anything else, Paula, asked the counsellor, hoping to get her to analyze her experience for the group. How did it feel to take charge of the phone bill, take charge of the chemicals, not have to ask for help? Now that you mention it, maybe that was the best part about it. Maybe it was.

The conversation never faltered, but the psychologist and the math teacher had more to do. Throughout the discussion the psychologist readily admitted her math anxiety, and the math teacher said things like "I make computational errors, too. That's why I need the red flagging that multiples of nine give me. I need it for myself." When the day's "lesson" (on tipping) began, it was the psychologist who went to the flip chart and started to explain. In half an hour, all manner of ways to calculate 15 percent had been explored. Funny stories were exchanged and even Marilyn had learned one sure way to compute a 15 percent tip. In her state, where the sales tax is 4 percent, simply multiplying the tax, which is usually written down on the bill, by four would produce a reasonable tip of 16 percent. Someone admitted during the discussion, "I never care what I eat, I just leave a dollar." Someone else told how she carries a pad and pencil these days and refuses to be embarrassed about having to calculate that way. Some slide to the nearest dollar or fifty cents to avoid complicated arithmetic. Others go overboard, multiplying the total by the decimal .15. In the discussion, of course, as people were explaining their methods, a fair amount of multiplication and division was taking place, but it seemed painless.

The assignment for next week: try to remember how your parents reacted when you had trouble learning math. Did it vary by parent? Did it vary by subject? And keep monitoring the math you do and the math you avoid.

Group Process

Some of the activity in this math anxiety workshop will be recognized by counsellors and psychologists as no more than good group therapy. The young women had selected themselves for this experience, but they had also been carefully screened for homogeneity of math background and of level of anxiety. They already felt comfortable with one another in this third session because during the first two they had shared their math autobiographies, tearing away the usual protections and avoidance mechanisms they had developed to deal with math-related issues. They were learning math. But they were also letting their feelings show and dealing openly with them. The dynamic between the psychologist and the mathematician was important, too. The mathematician revealed his own frailties and the counsellor showed that she was not very different, only a little more experienced, than the students. She kept the math teacher from moving too fast. In turn, he sometimes picked up feelings in the group that the counsellor had missed.

During the weeks between sessions, these young women were learning to recognize their feelings about math and to take some risks using math. They laughed a lot during the session I observed, and they gave one another considerable support. Everyone wanted Marilyn to understand and by the end of the hour and a half, Marilyn was smiling too. Some of the statements they made were interesting in themselves: "I don't feel anxious," one of them said at one point, "when I cannot balance the checkbook. I just feel dumb." "I'm not anxious at this point," another said while we were casting out nines. "I'm just not interested in what you are doing." These comments were also "processed," as group therapists say. Everything became grist for the mill of understanding feelings. Nothing was irrelevant. Everything was used.

In Middletown, Connecticut, Bonnie Donady, a counsellor, and Jean Smith, a math instructor, begin their adult sessions sharing math autobiographies. Out of this discussion comes a survey of what caused the group members to decide to overcome their handicap. The group meetings do not follow a rigid schedule. Donady and Smith assign "psychological homework," having the group members focus on the feelings aroused when they work on math. The first assignment might be to observe how they treat themselves when they encounter difficulty. Do they blame themselves? Castigate themselves? Or are they patient and self-reassuring? Another assignment will be to watch how they handle frustrations when doing math in comparison (usually in contrast) to how they handle difficulty in nonmathematical situations. Eventually, mathematics is introduced into the group session and the members are urged to monitor their own and others' reactions as they move from a meeting format into a class format. As soon as anyone begins to fall into a pattern of "succumb and surrender," the group leaders if not the group itself intervene.

Other places use a more structured agenda. At the University of Minnesota's math clinic, Sandra Davis sets up an initial interview, tracks clients into one of three different math classes, and then has them meet as a "support group" once a week for seven weeks during the semester. In one of those sessions, she concentrates on talking about math anxiety. In another, she does some assertiveness training and directs this training explicitly to handling the intimidation that takes place in math class. In a third, she engages the class in a behavioral desensitization exercise, having them relax all the muscles in their bodies and then asking them to concentrate on doing some math while they are relaxed.

At Mind over Math in New York City, Stanley Kogelman and Joseph Warren, both mathematicians and one a social worker as well, prefer not to do any math at all in their group sessions, concentrating only on the avoidance behaviors. They believe that once math anxious people come to grips with their feelings about math, they can learn anything. One of their clients, an adult man who had never been able to master long division, was disappointed at first that the instructors in Mind over Math refused to teach it to him. Yet one Monday night, at the group session, he proudly announced that his brother-in-law had taught him long division the previous weekend. The group surmised that his brother-in-law had probably tried many times before, but this time it had clicked because the man was more relaxed about math and more confident. The group sessions had primed him to learn math.

If the problem for many people in learning math is translating something they do not understand into language that makes sense to them, the presence of other people much like themselves will help immeasurably. In the course of a discussion about fractions or ratios or tipping, someone in the group may use a term a math instructor might not have thought to employ in this connection. This term will raise the curtain and the issue will make sense to someone else. I have seen this happen when, in a group of adult women, one complained that she had always had trouble turning feet into inches and inches into feet because she never knew whether the new amount should be larger than the old one. She was helped on the spot by someone else in the group, who suggested that before she undertake any measurement computation of this sort she should note whether she is going "from many to fewer" or "from fewer to many." "Many to fewer, fewer to many," the first woman kept saying to herself. She understood, possibly for the first time, how to cope with her disability.

I have been learning math in recent months via a process I can only describe as dialectical. Since the usual presentation in a math text often makes no sense to me, I try to write down in prose what I think does make sense. Then I interview mathematicians or people who know more math than I and ask them whether what I have written is correct. Sometimes it is and sometimes it isn't; but almost always I am told my question seemed to come at the issue sideways—sideways from their perspective, of course, not sideways from mine. Trying to write out something as it would make sense to you is one way to enter into that dialectical process. It is by no means efficient, in terms of time alone. But if it works it is the only way to begin.

There is no one right way to run a math therapy or math desensitization group. Nor does the mathematics discussed need to be as elementary as in the group just described. Generally it is advisable to have a psychologist and a math instructor share responsibility so that errors in logical thinking and mistakes in computation can be cleared up immediately. Most important is to eliminate anxiety-producing experiences: no tests, of course, no pressure to get right answers, no competition with classmates, no put-downs. This reversing of the usual mathematical experience may finally give the learner one positive math experience that will go far toward reducing anxiety.

Some people need no more than one positive experience. At Wellesley College, unusual instructional methods and an interesting set of problems seem to reduce fear and hostility toward math. Each problem area is first presented as a reasoning task, without numbers attached and without any formula given. The class, as a group, is invited to try out any number of approaches. If the unit is successful, the group will figure out a useful formula for itself. Even without group therapy or processing techniques, such a setting can evoke good feelings and reduce math anxiety.

At a community college in St. Louis, a math anxiety workshop begins with some guerrilla theater. The math instructor walks into the class on the first evening, turns sullenly to the blackboard, and begins to write some math examples. Although her demeanor is exaggerated, most of the people in the room take what she is doing quite seriously and wait expectantly for her to put them on the spot. A few minutes into the session, she throws over her authoritarian role and asks the class to talk about how she made them feel.

The assumption underlying all these different methods is that if anxieties are learned they can be unlearned. And so they can.

Talking about Math

People who don't like math don't like to talk about math. Part of their avoidance mechanism is to pretend that it does not exist. But math does not go away. People need it at work, in calculating percentages, in dining out, and in handling money. A math clinic is designed to integrate talking about math into the learning process.

Most educators know that feelings influence learning, but few take time out of class to deal with feelings. They may think that such a discussion is a distraction from the work at hand. One school in New York City experimented with a math clinic and had quite a different experience. High school students were taken out of their regular math class for one period each week (one out of three regular meetings of the class) to meet with two math therapists. The subject of the special class was feelings about math and the consultants intentionally did not teach any math at all. At first, the teacher was afraid that her class would fall behind because her students were missing one-third of their lessons. Soon she discovered, however, that her students were progressing more rapidly in the two classes they still had with her than they had previously done in three classes per week. Why? Because they were conquering some of the negative feelings that were keeping them from mastering math.

I believe that this talking process is at the heart of the treatment of math anxiety. As we have seen, it helps some people to know that they are not the only ones to suffer from fears of inadequacy about math or science. Moreover, the process of recollection, as stimulated by the math autobiography, can remove old obstacles to learning and provide insight into what is blocking learning now. At first, the participants talk either alone with a counsellor or in a group. Later the talking is transferred to the more private medium of the math diary or the tape recorder, though talking with others may continue. The eventual goal is to understand so well how we get in our own way that we can stop resisting and get down to learning math.

One caveat must be stated. Little is gained by taking people who are worried about not doing well in math and simply making them less worried about it. To tell them only that their answer is a little bit wrong and they should not feel bad about it is not much more than solace. Solace may be necessary, but by itself it won't cure math anxiety. Mathematics mastery is what we're after.

The test of success for any math anxiety management scheme will be whether one bites the bullet and signs up for an appropriate course in math, sticks with it, and does well. Since the anxious person on the road to recovery is still well outside his comfort zone, he might better take that course at a nearby community college than at the university, where he would compete with people who are very comfortable with math.

Some community colleges have recently restructured their courses in elementary mathematics to appeal to adults and meet their needs. Arithmetic and algebra are offered as "arithmetic techniques" and "algebraic techniques" and the curriculum advances slowly but surely through "linear mathematics" and "elementary functions" or "probability" to two terms of precalculus. In all, students can profitably spend six semesters getting ready for the calculus or statistics. One does not have to try to learn math at the kind of institution or the pace at which one might study history or literature. The community college, with its emphasis on teaching and its careful, nonthreatening curriculum, may be just the place for a math avoidant adult—even someone with a previously earned B.A.—to begin to rebuild her confidence and skills.

Diagnosis

The first stage in any treatment process is diagnosis. In math therapy this takes the form of an initial interview, in which the counsellor asks about past experiences in math. Out of this questioning comes what we call the mathematics autobiography. How well do you think you understand arithmetic, algebra, and geometry now? What feelings have you been harboring about mathematics? What kind of schooling did you have? Who helped at home? Do you remember what people said to you when you made a mistake in math or when you did unexpectedly well? When did failure begin? Do you feel that you don't know how to read a math textbook? Do you fear that you are slow? When you solve a problem, do you know why? Do you think that you have been away from math for so long that you have forgotten everything?

There are some tests of math anxiety that can be purchased and administered at home, but nothing replaces a well-structured interview. An initial diagnostic interview might go something like this:*

INTERVIEW 1

Q. Why don't you begin by telling me about your math experiences.

A. In elementary school I was always in the top math group although I could never figure out why. Then in fifth grade I got a C on a test. That had never happened before. I began going to my father for help with my math.

Q. Was this for after-school help?

A. Yes, and whenever I was confused and felt that I didn't know it, I would wait until I was home and ask my Dad.

Q. Couldn't your teacher help you?

A. I guess she could have, but I must have felt better asking

*The following examples are composites from interviews conducted by Bonnie Donady. Commentary was prepared jointly by Ms. Donady and the author.

my Dad. I never wanted to appear stupid, and when I worked with my father I could get it. In class, and under pressure, I couldn't do it.

Q. How was math as you continued in school?

A. I guess as I think about it I just continued going to my father. In seventh and eighth grades I understood more, but still had a sense of being in the dark. Particularly in eighth grade, that teacher was antagonistic to questions. Sometimes even with my Dad I had a sense that I didn't have it all together. I was always sure that something was missing.

Q. How was your high school experience?

A. I did fine. I liked geometry and got 98 on the algebra regents, but I was never as good as the math whizzes. I just couldn't work fast enough.

Q. What happened in calculus?

A. I had a typical teacher who seemed to explain math in the hardest way.

Q. Did you ask questions in class?

A. No, I never asked questions in math class at all after fifth grade.

Q. Did you ask questions in other classes?

A. Yes, but not in math class. I never wanted to appear stupid and I didn't want anyone to know I was confused. So I would wait and ask my father. And now I miss having him to ask those questions.

Q. What course are you taking now?

A. I'm taking the calculus course and having trouble with the algebra.

Q. Are you certain it's the algebra? It seems to me that you need a safe place to ask questions. Can you ask the Math Clinic tutor questions?

A. Sometimes, but not always.

Q. What happens when you do not understand something?

A. I begin to feel uncomfortable, and I wait until after class to work on it. My mind doesn't seem to consolidate the information or see connections in class.

Q. Do you think other people have the understanding?

A. Well, no one asks the questions so I guess they do.

Q. Perhaps they, too, are fearful of risking embarrassment?

A. I guess so.

Q. What do you plan to do?

A. I want to finish calculus, and probably take linear algebra next year. My father was a math major and he would like me to continue taking more math. He says it will always be useful.

Q. How do you feel about it?

A. I know that I will feel successful if I complete these courses. Sometimes when I don't understand it, though, I feel like stopping.

It is interesting to note how important this girl's father is in her math autobiography. She doesn't want to disappoint him. She doesn't want to disappoint her teacher either. She fears appearing too stupid in class. Like many other girls (and boys) she did not take advantage of her teacher's willingness to answer questions, because she never asked them. Her problem was not math; it was herself, her pride, her need to appear smart, and her impatience. This is the kind of insight the interview often provides.

INTERVIEW 2

Q. What are your reasons for coming to the clinic?

A. I know that I now have to do something about my math. I want to go to college now and I have to take the SATs before applying to college. I have been out of high school for two years.

Q. How do you feel about doing some math now?

A. Terrified. I have not done any math in four years and I am scared to try again.

Q. Is going to college worth taking math again?

A. I really am not sure, but I am determined to at least try it.

Q. That sounds very good. What is your math background like?

A. I stopped taking math in my sophomore year in high school after a really bad experience with geometry. I swore that I would never take any math again and here I am considering it!

Q. What happened in geometry?

A. I just hated memorizing all those theorems and even though I had done well in algebra I was failing geometry.

Q. There are two important points here, first that geometry was your first negative math experience—is that right?

A. Yes. I did fine in elementary school and also in junior high school, and as I said I had an A in algebra.

Q. I really want to know more about how you felt in geometry, but first I'd like to know if anyone suggested that many people have one bad math experience, but still can continue. I know a mathematician who failed seventh and eighth grade math.

A. Well, no one suggested that I try any more math. In fact, my mother said it was okay to stop, because she had never done well in math either.

Q. Let's keep in mind that geometry was your first and only negative math experience, as you tell me a little bit more about that situation. For example, did your difficulty focus on any one part of geometry—tests, class discussion, homework?

A. I don't remember; the whole thing is such a blur I can just remember feeling like a dummy, and feeling as if I had just reached that point in math that I should stop. Somehow I felt that if I couldn't do geometry the rest would also be hopeless.

Q. Did everyone who had difficulty with geometry also drop math?

A. I don't know. I never thought about that. I was so successful in my other courses it seemed masochistic to continue with something causing me pain.

Q. Now that you have reflected about it, how do you feel?

A. I guess that I am sorry that I didn't do something else.

Q. I think we might talk some more about your geometry experiences, and in addition it might be helpful if we did a little math so that we could plan the appropriate level course for you to take. How does that sound?

A. That sounds good.

Here is a student who let one bad experience get her down. She probably had some bad experiences in other subjects, too. Why did she single out math as her failure subject? Some people remember early failure in arithmetic as the result of having missed some crucial explanation while they were out sick. Years later they will say that they never understood subtraction, for example, because they had the measles when it was being taught. This may be. But the counsellor is suspicious, because they also missed spelling and reading and geography during those weeks and do not experience later failure in these subjects. What is there about math, or is it something about the attitude students bring to math, that makes them feel so hopeless about making up missed work?

Many people think that because math is a cumulative subject, previously missed material can never be made up. Sometimes this is so, but not always.

Note that at the end of this interview, the student feels "good." This is the first step in rebuilding confidence in learning math.

INTERVIEW 3

Sometimes it is better for a person to stop taking math for a while. The interview can help her find this out. If this is best, then once the decision is made the important thing is not to feel guilty about it, because all decisions can be altered.

A. I was always told that everyone knows how to do math and that I was just not applying myself.

Q. Did you think you weren't applying yourself?

A. I knew I just couldn't get it and I did badly in math in sixth, seventh and eighth grade. I took just one general math class in high school and did badly in that too. Then I went into the service and did really well (As) in everything except medical math.

Q. What happened then?

A. Well, I was publicly ridiculed repeatedly and I just felt awful so I began avoiding anything that looked like it was vaguely associated with math.

Q. How is it in your present math class?

A. Well, the teacher is better. She tries so hard and really wants me to get it.

Q. You don't want to disappoint her?

A. Right. That's why I even came here, but I just can't do it. But I know if I can't do math I just never will be able to own my own business.

Q. Do you know enough math now to do what you need to do?

A. I doubt it, but maybe I can do my art.

Q. Is it worth it to take math or should you focus on your art?

A. I don't know. I just feel so terrible when I'm in math class.

Q. How?

A. Stupid. Like I'm a moron, no good at anything. I just hate myself.

Q. What should you do then?

A. I'm not sure.

Q. How about trying a kind of game? When you sit in the blue chair you'll be taking math. When you sit in the red chair you won't. In each chair talk about how you feel. Let it flow.

A. Okay. I'll sit in the red chair—I just don't want to sit over there yet. I know I'm okay. I can do art, I know a lot about good health. Why should I make myself miserable taking math?

Q. Come over here and answer that.

A. *(in the blue chair)* I should take math so I can own the farm. Everyone should be able to do basic math. Only idiots can't handle numbers. *(At this point she broke into tears and was unable to talk. Once composed she quickly got back into the red chair. When asked where she felt more comfortable, she said it was in the red chair not doing math but she wasn't sure she could allow herself that freedom. We continued talking; occasionally she'd sit in the blue chair, doing math. She would repeatedly start sentences with "I should." In the red chair her posture was different. She sat up and felt more sure of herself, but she was sad.)*

A. I guess I won't do any math now. I'll drop this course and see how I feel about it in a while.

Q. How do you feel now that you've decided?

A. Relieved alittle, lighter, sad. Maybe I'll change in some months.

The interview proceeds from math history into a chair game. Moving from the red chair to the blue chair and back again is a way to deal with two conflicting feelings: one a positive interest, the other a feeling of inferiority. In this case, the girl was so upset about math that the counsellor advised her not to elect any math courses, at least not for a while.

Math anxiety is not limited to females. Many males often find math difficult or painful too. But when a male suffers from math anxiety he may present his problem differently, because he is still trying to have a "manly" image. Boys, in general, believe that if they work hard they will succeed, and if they fail it is because they had a bad teacher or a bad book, not because they have a nonmathematical mind. This helps many boys to continue in math in spite of one bad experience. The next interview is with a boy having trouble deciding whether to take more math.

INTERVIEW 4

Q. How do you feel about math now?

A. Math is not such a big deal. I'm not so impressed with people who can do math. And I'm definitely more confident because I can do word problems now!

Q. What are your plans now?

A. Well, I'd like to take self-paced calculus, but I have to pass every course I take next year since I had to drop economics this year. I'd like to take that same teacher and show him that I can do it.

Q. What about calculus?

A. It sure would be terrific if I could do it but when the teacher gave me a problem to do, he said I'd have difficulty because of the way I worked the problem.

Q. How did that make you feel?

A. Just like in high school—inadequate, really not sure I could do calculus.

Q. Let's try a game. When you sit in this blue chair you are the person who can and will take calculus. When you sit in the red chair, you don't take it. Okay?

A. *(blue chair)* I want to take calculus. I could really use it, I'll be proud of myself, and I'll feel so great when I finish it. It'll allow me to do more things.
(red chair) But what if I don't finish it? I have to pass every course. If I take it and fail or have to drop it I won't graduate. *(His posture is slouching.)*

Q. Come over here and answer that.

A. *(blue chair)* I can finish it. Steve said I can do it and I finished the algebra and can now do economics and really feel good. Besides, I can start it this summer.
(red chair) I just don't know. Algebra is one thing, but calculus—that's really hard stuff. What if the instructor is right and I can't do it?
(blue chair) I'll get tutoring. I'll work at it and I'll do it.

Q. How are you beginning to feel? Do you notice you sit and talk differently in each chair?

A. Yes, I guess so—I'm really more comfortable in the blue chair.

Q. What is your decision then?

A. I'm going to take it. I'll start it this summer and I'll sign up in the fall.

This boy needed to review what had happened in the math course he had just completed to reassure himself that when he continued in math he would still succeed. The chair game helped him see that he could be comfortable taking more math, because he felt better in the blue chair than he did in the red chair.

If we can learn to diagnose ourselves, then we can free ourselves to some extent of the support and guidance a counsellor provides. The math diary, described in Chapter Two, is such an extension of diagnosis. It is also a way to monitor one's progress:

Today was a day filled with math. First I had to figure out the tip. Boy, I began to feel my palms sweat and my mouth went dry. We didn't ask for separate checks and we had to divide it all up. I was really tempted to say I'd pay for the whole thing so we could only get out of there. But I didn't. I kept telling myself it is okay to feel nervous. I could see that my hands were dripping wet but I figured that if I did it slowly I could figure out which part was mine. I noticed that I began telling myself that I was a dummy for needing a pencil and paper, but I used one anyway and I figured it out correctly. Later on, reading the newspaper, I noticed that there were some graphs and charts on the page I was about to skip. I decided for once not to skip that part and instead to try to understand what they were saying. Again the dry mouth and the sweaty palms. But that's simply what happens to me when I get nervous.

It is possible, though not desirable, to do self-diagnosis. By completing the following sentence questionnaire, one can begin to analyze one's feelings toward math. This "Sentence Completion Questionnaire" was designed by Sandra L. Davis for use in the continuing education math class she directs at the University of Minnesota.

Finish the sentences in a few words or paragraphs and then look over the picture you have drawn of your attitudes toward yourself and toward mathematics.

1. When I got bad grades in math ____________
2. I did well in math until ____________
3. My high school math teachers ____________
4. My father felt that math was ____________
5. Math is ____________
6. When it was time for math in grade school I ____________
7. My ability to do math is ____________
8. My grade school math teachers ____________
9. I want to become better at math so that I ____________
10. When it comes to math boys are ____________
11. My mother felt that math was ____________

12. When it was time for math in high school I ________
13. My background in math is ________
14. Doing math makes me feel________
15. People who are good at math ________
16. I liked math until ________
17. When it comes to math girls are ________
18. My most positive experience with math was when ____
19. When it comes to math, I find it difficult to ________
20. Female mathematicians are ________
21. I haven't done math since ________
22. If I were better at math I would ________
23. My most negative experience with math was when __
24. When I got good grades in math ________
25. My motivation to do math ________
26. When I hear someone say "math is fun" I________
27. Male mathematicians are________
28. I am anxious about math because________

Immersion

> When asked how he had discovered the universal law of gravitation, Newton answered that it was because he thought about the problem all the time.
>
> —Anonymous

If, as some have claimed, mathematics anxiety is caused by nothing more than inadequate math preparation, then one way to overcome it is to get back into math at the appropriate level as soon as possible. Some will be deterred by fear, believing that since they stopped at, say, the sixth grade level, it would take them seven to nine years to reach the point where they could tackle the statistics and calculus they need. Not so. With the sophisticated mental equipment we have as adults, sharpened by years of experience in learning, we can make up time extremely fast.

Alan Natapoff is a trained physicist and a researcher of the structure of the brain. He believes and has demonstrated that adults can be taught arithmetic, algebra, and even introduced to the ideas of the calculus in a matter of weeks, if appropriate procedures are employed. Natapoff has used his techniques with learning disabled retarded children and with adults in several locations. He bases his theory on the belief discussed in our treatment of word problem solving, namely that the part of the brain that thinks best (creatively and abstractly) is extremely inefficient in other respects. Because of this, Natapoff argues, it may take a hundred times longer to learn one complex idea than it would take to learn it if it were broken up into ten more quickly assimilable "bits."

Obviously Natapoff thinks that math is best approached at this level not as a subject to be comprehended but as though it were a set of skills to be mastered. He believes that if such skills can be rehearsed somewhat like bike riding, ping-pong playing, or choral singing skills, they can readily be mastered and will provide an important sense of comfort and certainty to the learner. Math, from this perspective, is not just a collection of ideas; it is also something people do. And to learn how to do it well and confidently, one must do it often and intensely.

To teach math skills as we learn singing, batting a ball, and even as we once learned to walk, Natapoff selects a task and breaks it up into what he calls "rehearsable bits." Rehearsing math means using the relevant sensory apparatus, listening to phrases repeated out loud, speaking these in unison, touching materials, writing on the board (as a group, not just one by one), and so on. Only when the bits are really learned will they be put together into longer chains of complex performances. Too often, Natapoff argues, a concept is presented in math class only abstractly and the learner never gets a chance to feel it and to make it part of the brain's repertoire. Natapoff believes that our short-term memory is crucial for creative abstraction but it is flawed in that it readily forgets.

As an example of what he is describing, Natapoff refers to the algebraic concept of substitution. Part of solving an algebraic equation may involve substituting an expression like $(x+3)$ for the expression y. Instead of simply explaining this, Natapoff will have his students, even adults, say out loud "*parenthesis* $x+3$" every time he says "y". This is not rote in the old-fashioned sense, he insists. It is an attempt to etch the idea that every time I say "y" you say "*parenthesis* $x+3$" into the student's brain until, through practice, and because each sequential step seems so easy, fear disappears and the idea of substitution is learned.

Natapoff has success with adult learners. I have watched him bring three different groups of adult women forward through arithmetic, algebra, and pre-calculus. Maybe he is just another magical teacher, but at least in his view there is biological evidence that certain skills, to be truly learned, must be learned with more of the body than we ordinarily use in the classroom. An analogy to his immersion of students in arithmetic can be taken from choral singing. People learn to sing by doing many things at once: hearing the music as sung (or played) by others; reading notes with the eyes and brain; feeling the rhythm of the music; and even, in previous decades, by having differently shaped notes represent different sounds so that the visual faculty that discriminates shape could also be brought to bear on the activity of reading music.

We don't have to have Natapoff as a teacher to use some of his ideas on our own. We can "practice math" by increasing our awareness of numbers and playing with their relationships. We can devise our own verbal, visual, aural, or tactile modes of mastering mathematical ideas.* Above all, Natapoff's demonstration of how fast we can refresh our knowledge of arithmetic and algebra, in a weekend of intensive work, for example, should encourage us to plow ahead.

Most mathematics learning, Jerome Bruner writes, requires "reordering, unmasking, simplification and other activities, akin to the activity of a speaker, and not a listener." The passive learner will not do as well as the active learner. How then to become an active learner on our own?

On Your Own

Pre-Season Training

If some part of mastering math is comparable to learning to walk or ride a bicycle, then that part ought to respond to training, exercise, and frequent practice. If I were going to learn tennis today, I would not take a racket in hand until I had done six months of running and lifting weights. I did not know it when I began trying to learn the game, but most of my difficulties had

*A. Binet, the psychologist famous for co-producing the most widely used intelligence test, once called for "mental orthopedics" to achieve the same thing.

to do with speed on my feet, endurance, and power in my arms and legs. It seems to me now that had I gotten into top physical form first and then been given a racket, I might have done a whole lot better than I have. The same may be true for mastering mathematics. It calls for muscular, visual, and cognitive coordination that we can exercise in advance. Part of our math anxiety as adults comes from having rusty faculties.

Three areas suggest themselves as suitable for training and practice: spatial skills, number play, and puzzle solving. Spatial skills can be approached in two ways. The first is simply to learn how to solve those figure problems that are given on tests. Paul Jacobs, a specialist in psychological testing, has written a book with numerous examples of two-dimensional figure problems and hints on how to solve them.[1] Although this type of exercise might first appear altogether too test-oriented, the fact is that doing enough of these figure problems under the author's supervision does increase skills in spatial visualization. The same is true for certain kinds of spatial games.

The second approach involves using materials that help us visualize mathematical relations. Many of these materials did not exist or were not widely used when I was in school. The Montessori method has long used concrete materials to illustrate the number line and other kinds of numerical and fractional relationships. Although high school students and adults might incorrectly think of these materials as beneath them, Persis Herold, author of *The Math Teaching Handbook,* recommends them highly for all ages.[2] These materials help make palpable the meaning of a fractional part, an area, even a change in one variable.

One good example is the geoboard, a simple rectangle usually made of wood, dotted with pegs or nails in some regular order (like the stars on the American flag). Using rubber bands, which can be stretched around two or more nails to fashion some kind of polygon, we can create any number of shapes and begin to see quite realistically how area relates to perimeter and what happens when the perimeter changes. Area seems quite tangible on a geoboard: it is not just some arbitrary manipulation of length times width. It is easy to underestimate the value of such an experience, for the real excitement in mathematics is in the abstractions. But if we actually get our hands on these elements we are more likely to understand and hence remember what area is all about. Jane S. Stein at Duke University's Continuing Education Program, as well as Persis Herold, has used geoboards, tangrams, and other "elementary school materials" quite successfully in a program for adult women.

Dotted paper, sold by some companies but easily made at home, can do almost as well as the geoboard. One can draw lines connecting the dots to form shapes. Graph paper is not obsolete either. By graphing functions, we can get used to what happens when measurements are compared.

Once upon a time bisecting an angle or "dropping a perpendicular" in geometry class was an exercise in frustration. I was taught to use a compass and a straight edge. One of the reasons I never appreciated that the lines bisecting the three angles of a triangle meet in the middle at a single point is that my bisections never did meet. I was a messy, inaccurate draftsperson, then as now. But with a newly devised miracle tool, appropriately named a Mira, bisecting an angle and transforming a geometric shape is a snap.

The Mira is a single plastic surface that is both transparent and reflecting.[3]

Number play might take several forms. Since getting involved again with mathematics, I have found it useful to expand my multiplication table up through the 15s. This makes me feel secure in estimating many two-digit multiplications. To know that $15 \times 7 = 105$ quite as readily as to know that $6 \times 7 = 42$ gives me a better sense of the size of things. Another good group of numbers to memorize are the most common fractions and their percent approximations: 8 percent is roughly the same as $1/12$; 6 percent is roughly the same as $1/16$. These come in handy for making rough percentage estimates and for going quickly from percents to fractions and back again.

Fraction	*Percent* (approximate in some cases)
$1/2$	50 percent
$1/3$	33.3 percent
$1/4$	25 percent
$1/5$	20 percent
$1/6$	16.7 percent
$1/7$	14 percent
$1/8$	12.5 percent
$1/9$	11 percent
$1/10$	10 percent
$1/11$	9 percent
$1/12$	8.3 percent

It is not always necessary to memorize these numerical equivalents. Like the man who carries a multiplication table around in his wallet, we too can create sets of tables, formulas, or even problem-solving hints on cards that we carry around. No one expects us to figure out what day of the week it is from the date and year; we are supposed to consult a calendar. In the same way, we can refer to our private tables. Soon this inventory might even slip quietly into our conscious memory.

Problem-solving ability is certainly improved by doing puzzles and games. Cryptarithmetic, which is deciphering an arithmetic operation written out in letters, is a particularly good mixture of problem solving and numerical play. Some examples of letter division, one form of cryptarithmetic, follow.

Letter Divisions

(Courtesy of Paul J. Tobias)

Note: Each letter in the long division stands for a digit from 0, 1, 2, 3 up to 9. The same letter stands for the same digit throughout the problem. ? is like a letter.

```
        F E A R
     ___________
OF / M A T H ?
     M R
     ---
       ? T H
       ? R T
       -----
         ? U ?
         ? F F
         -----
             A
```

SOLUTION

Look for some clues. The *E* in the quotient is evidently zero because the *? T* is smaller than the *OF* and we had to bring down another letter *H*. The *O* in the divisor is higher than the *?* (otherwise we wouldn't have had to bring down the *H*) and the *A* and the *T* are higher than the *R* (since we can always subtract successfully). The *T* is one more than the *A*. The puzzle solver begins to limit *F* to 2, 3, or 4 because if $F = 1$ then multiplying by *F* would produce the same results. *F* cannot be 5 or 6 either because 5 multiplied by 5 would produce a 5 (another *F*) and 6 multiplied by 6 would produce another *F*. One step further with *F*. If we assume that *F* is 3 then *F* times *F* would equal 9, but we have already decided that *A* is higher than *R*, which eliminates $F = 3$. (No number is higher than 9.) So *F* is now either 2 or 4. Trying $F = 4$, *O* can only be 2; otherwise the multiplication of *F* times *OF* would produce a three-digit number. If *O* is 2, then *?* has to be 1, *R* is 6, *M* is 9, and $A = 7$. *R* from $T = ?$. Since *T* cannot be 7, it must be 8. If *F* is 4 then *U* has to be 5 since the difference disappears when there is a carryover in the last subtraction. At this point *H* can be derived since *T* from $H = U$ and we know the values of *T* and *U*.

The Code:

0	*E*
1	*?*
2	*O*
3	*H*
4	*F*
5	*U*
6	*R*
7	*A*
8	*T*
9	*M*

Voilà!

Here are some others. (Answers on page 278.)

```
          B O Y
        ____________
GIRL ) S C H O O L
       R C B B
       -------
       L C S G O
       L Y G H O
       ---------
         B O G H L
         B I B B Y
         ---------
           L H Y I
```

```
                 S I X
            ________________
T H R E E ) E I G H T E E N
            E R G N H G
            -----------
              N I S X E E
              N Y I H I X
              -----------
                S X Y E T N
                E I H T X E
                -----------
                  N G X H Y
```

Successful problem solvers, we are told, play with a problem rather than immediately setting out to solve it. We who are less confident rush for the formula even before we really understand what is being asked for. The breakthroughs in understanding math come suddenly when we realize we have what we need to know in our heads if we can just relax and let our minds use it.

We do not know enough about the workings of the brain to say for certain what the best pre-season training might be. For some, problem solving may be the best practice. For others, just thinking about numbers and their relationships will help. For still others, reading in the history of mathematics can be least painful.

The instructors in the math clinic with which I am associated advise everyone to do word problems regularly every day. That's probably not a bad way to begin.

Reading Mathematics

Natapoff says that our math learning is not "oral" or "aural" enough and attempts to compensate for this in his remedial classes by having people speak out loud what they are being told. Another criticism of conventional math teaching is that we are not taught to read mathematics so that we can learn it on our own. Those of us who do not go on in math, it is true, never really learn to study mathematics the way we study other subjects. Studying math as we do it in the lower grades consists of review and drill. Studying on our own involves learning to understand new material without the help of a teacher. To do this effectively, we have to learn to read math, and that is what many of us do not know how to do.

One way to begin to read mathematics is to recognize how we read other subjects. Most texts in the humanities and even in the social sciences state important ideas and facts more than once. Therefore, those of us who enjoy reading this kind of material learn to read quickly, even to skim, to get the gist of what is intended, and we do not worry too much about missing something. Chances are if a fact or an interpretive statement is important it will be repeated or paraphrased. Topic sentences, paragraphs, the structure of a well-written essay are all signposts to tell the rapid reader where to slow down and even to stop. But reading mathematics is reading for immediate mastery. Things are stated only once and must be well understood before we move on. The process is so different from ordinary, even serious reading that the word "reading" might itself be misapplied.

Mathematicians tell me they would not think of tackling a math book without a pencil and paper. They try to sketch, if possible, what is being said. They stop and imagine examples that would illustrate the problem at hand. They ask themselves questions and try to answer them. Above all, they move slowly, very slowly over each part of each statement.

Relapse

> One barrier falls and the next one is higher.
>
> —Old Chinese proverb

Let's say you're on your way. You have gone through one set of math anxiety workshops and signed up for a review algebra course. You're doing puzzles and word problems every day and figuring out graphs and tips and percentages. You feel good about math and even better about yourself in general, and then one day, in class or out, you encounter a problem or procedure or quantitative analysis that is not just unfamiliar; it is impossible! Chances are all the old misgivings and "I can'ts" will come flooding back again and it will seem like sudden death.

The fact is that math is not easy; and that math anxiety is experienced not only by people like the person you used to be but by experts too; and it does not occur only at the elementary levels of arithmetic. Able students of math who succeed in high school get the feeling of sudden death doing calculus. Does this mean math anxiety cannot be "cured" or eliminated after all? Probably true. But, if it is also true that mathematicians too feel anxiety when confronting a problem of a sort they have never seen before, then the way out is not to deny the anxiety but to manage it. Self-talk, time out for careful consideration of the issue, going for help to someone who has been helpful in the past, can get you over that barrier.

In some cases the formerly math anxious will become so attached to a particular teacher, even to a particular text that finally helps them make sense of the material, that they have something akin to withdrawal symptoms when they have to move on to the next class or the next book. Psychiatrists who experience this in their patients' temporary dependence on the doctor call it a transference. Not yet entirely autonomous, the learner attributes her success to some extraneous element, the teacher or the book, instead of to herself. An apparent relapse, then, might be no more than an old response to unfamiliar terrain.

It is important to go through this at least once because managing anxiety is just that: experiencing anxiety and mastering it. Besides, given the inertia in the school system, adults as well as children will probably have to cope with imperfect teachers and texts for a long time to come. Competence only in the ideal, protected situation is not enough. The real goal is autonomy.

Appendix I

The following is a composite of confidence and anxiety rating scales that have been developed by Dr. Elizabeth Fennema and Dr. Julia Sherman. A full set of these scales is available for $5.00 from the American Psychological Association, 1200 17th St. N.W. Washington D. C. They are called the "Fennema-Sherman Scales." The full set includes the following:

1. Attitude Toward Success
2. Math Anxiety
3. Confidence in Learning Math
4. Usefulness of Math
5. Perception of Mother's Attitude
6. Perception of Father's Attitude
7. Perception of Teacher's Attitude
8. Affectance Motivation

Composite Math Anxiety Scale

For each statement give a number 1–5 which indicates whether you strongly agree (1) or strongly disagree (5).

I usually have been at ease in math classes.
I see mathematics as a subject I will rarely use.
I'm no good in math.
People would think I was some kind of a grind if I got As in math.
Generally, I have felt secure about attempting mathematics.
I'll need mathematics for my future work.

I'd be happy to get good grades in math.
I don't think I could do advanced mathematics.
It wouldn't bother me at all to take more math courses.
For some reason, even though I study, math seems unusually hard for me.
I will use mathematics in many ways as an adult.
It would make people like me less if I were a really good math student.
My mind goes blank and I am unable to think clearly when working in mathematics.
Knowing mathematics will help me earn a living.
If I got the highest grade in math I'd prefer no one knew.
Math has been my worst subject.
I think I could handle more difficult mathematics.
Winning a prize in mathematics would make me feel unpleasantly conspicuous.
I'm not the type to do well in math.
Math doesn't scare me at all.

Appendix II

Programs for Math Anxious People

Wesleyan University Middletown, Connecticut.

Math Clinic for undergraduates. Six-session workshops for adults run by Bonnie Donady, counsellor, and Jean Smith, math instructor.

Concord, Massachusetts.

Mathematics Learning, Inc. Directed by Michael Nelson, M.D., psychiatrist, and Deborah Hughes-Hallett, math instructor. Twelve-week sessions for adults.

New York City.

Mind over Math, Inc. Sessions for math anxious adults. Directors are Stanley Kogelman, math Ph.D. and certified social worker, and Joseph Warren, math Ph.D.

University of Minnesota, Minneapolis-St. Paul, Continuing Education for Women

Counsellor Sandra Davis has perfected a six-session model in which materials are read, test anxiety is dealt with, assertiveness training is given, math games are played, and self-defeating behavior is studied.

Iowa State University at Ames, Iowa, 1976–77.

Jacqueline Pedersen, mathematician, uses a model including behavior modification techniques. She includes the verbalized math autobiography.

University of Indiana, South Bend, Indiana.

Math Without Fear sessions run by William Frascella, chairman of the math department.

Oswego, New York.

Fred Fischer, professor of math at SUNY Oswego, runs a workshop on math anxiety for the adult education system.

Hood College, Frederick, Maryland.

Classes in math for math anxious undergraduates.

University College, Syracuse, New York.

Classes in math for math anxious undergraduates.

Schoolcraft College, Livonia, Michigan.

Shirley Emerson, counsellor, and Barbara Riehl, math instructor, offer math anxiety workshops for adults.

Duke University, Continuing Education, Durham, N.C.

A systems analyst and a counsellor run a Math Laboratory for adults using concrete materials to abate unfamiliarity and anxiety.

Sarah Lawrence College, Bronxville, New York.

Mind over Math desensitization program plus two courses in review algebra and an introductory course in statistics constitute a program for returning adults who are math anxious.

California State College at Long Beach, California.

A program of workshops offered by Ruth Afflack, of the math department.

Mt. Vernon College, Washington, D.C.

"Math for the Wary," a continuing education course taught by Persis Herold.

Indiana University at Bloomington.

Math instructor Darlene Keihn and Counsellor Gayel Stuebe run a continuing education seminar for men and women dealing with: strategies for basic facts, math symbols, statistics, graphing, problem solving, and use of the calculator. The program also includes individual sessions.

San Diego State University.

Gail Small at SDSU using undergraduate students in mathematics provides non-credit help discussion groups on math, a mini math course, and weekly sessions of math games for persons who are fearful of math.

University of California, San Diego.

Marilyn K. Simon offers a course in "mathphobia" for adults during the summer as part of the extension program. Simon, a professor in the department of math education, also runs a course for elementary age students (mostly girls) with math anxiety. She calls this course "Math Power."

University of Wisconsin, Madison (Extension).

Laurie Reyes offers a course for the math anxious as part of the Extension summer program.

Elizabeth Fennema is producing "Math Education Transportable Visual Modules". Department of Mathematics Education, University of Wisconsin, Madison Wisc.

Other Types of Activities

The following do not focus directly on math anxiety but on math avoidance in young women.

Wellesley College, Wellesley, Massachusetts.

A discovery course in elementary mathematics under the direction of mathematician Alice Schafer offers new kinds of material for college women.

Mills College, Oakland, California.

Under the direction of mathematician Lenore Blum, an innovative curriculum together with peer counselling has trebled the mathematics enrollment at this women's college.

Lawrence Hall of Science, Berkeley, California.

A program of lectures and discussions about occupations in math and science for junior high school girls, under the direction of Nancy Kreinberg.

References

1. Paul Jacobs, *Up the I.Q.* (Conn., Wyden Press, 1977).
2. Persis Herold, *The Math Teaching Handbook,* 1978. Available from S.E.E., 3 Bridge St., Newton, Mass.
3. The MIRA and its uses are the subject of a workbook by Iris Mack Dayoub and Johnny W. Lott. Privately published by the authors and available from Iris M. Dayoub, 12508 Over Ridge Rd., Potomac Md. The MIRA is available from Cuisenaire Company of America, 12 Church St., New Rochelle, N.Y.

Answers

Code for *GIRL SCHOOL BOY*

.0. H
.1. L
.2. B
.3. G
.4. I
.5. O
.6. R
.7. Y
.8. S
.9. C

In *GIRL SCHOOL BOY* the *H* is definitely zero—*O* minus *O* is *H* and *G* minus *H* is *G*. Then *C* minus *C* with a carryover of 1 equals *C* equals 9. *S* is 2 over *R*. *G* and *B* are limited to 2 or 3, but which to which?

Code for *THREE EIGHTEEN SIX*

.0. G
.1. R
.2. H
.3. Y
.4. E
.5. S
.6. X
.7. N
.8. T
.9. I

In *THREE EIGHTEEN SIX* the *G* is definitely zero—note that *G* from *E* is *E*. Note also that *G* for zero appears twice in the subtrahend, which indicates that *S* is 5 and *E* and *H* are 4 and 2 respectively. When *G* is subtracted from *G* with a carryover 1, the remainder *I* becomes 9. Take it from there.